T5-ARX-010

Maintaining Esthetic Restorations

Lynn M. Miller R.D.H.

edited by

Michael B. Miller D.D.S.

■■■■■◆ ◆ ◆

MAINTAINING
ESTHETIC
RESTORATIONS
PAGE ii

Copyright © 1989 by Reality Publishing Company, Houston, Texas

All rights reserved. No part of this book may be reproduced or transmitted in any form or by any means, electronic or mechanical, including photocopying, recording or by any information storage and retrieval system without the written permission of the Publisher, except where permitted by law.

Printed in the U.S.A.

ISBN   0-9623707-0-3

ACKNOWLEDGEMENTS

The design and beautiful artwork was done by Dr. Joan Chamberlain. Not only is Joan a great dentist and beautiful artist, but one of the funniest, wittiest, people I have ever met. She has kept me going during the writing of this book to see it to the end.

Dr. Craig Mabrito has been my guiding light. Everyone needs someone who "tells it like it is", and Craig was there every step of the way.

Linda Nash R.D.H., took the time to contribute many vital and interesting points to the "Questions and Answers" section. She was a "Cosmetic Hygienist" before anyone else knew there was such a thing. Linda is now teaching her valuable hygiene skills to hygiene students, and her love for our profession is a true inspiration.

Lynn Lyons contributed a great deal to the "Questions and Answers". Lynn not only gave me questions, but answered many of mine. We graduated from different universities and Lynn gave me a new perspective. Her sincere desire to achieve the best for her patients and profession shows in everything she does.

And last, but certainly not least, thanks to my best friend, teacher, editor, and husband, Dr. Michael B. Miller.

This book is dedicated to my mother, Margaret Cosnahan, whose constant caring for everyone she met taught me what life is all about. This constant compassion and benevolence is essential to every good hygienist.

To my sons Heath and Chris, who constantly help me smile and laugh, especially at myself. Thank you Chris for giving me the time I needed to make this book come true.

◆ ◆ ◆ ▬▬▬▬▬

WHAT IS A SMILE . . .

It costs nothing but creates much. It enriches those who receive without impoverishing those who give. It happens in a flash, and the memory of it sometimes lasts forever. None are so rich they can get along without it, and none so poor they are not richer for its benefits. It creates happiness in the home, fosters goodwill in a business, and is the countersign of friends. It is rest to the weary, daylight to the discouraged, sunshine to the sad, and nature's best antidote for trouble. Yet it cannot be bought, begged, borrowed, or stolen, for it is something that is no earthly good to anyone until it is given away. And if in the course of the day some of your friends should be too tired to give you a smile, why don't you give them one of yours? For nobody needs a smile so much as those who have none left to give.

Dr. Larry B. Lanham, Grin and Share It

TABLE of CONTENTS

PREFACE

Our profession is in a time of turmoil. Dentists are upset and angry over the perceived notion that hygienists are trying to branch out, start their own practice, and establish an identity apart and separate from the dentist.

Realistically, most hygienists do not want to open their own offices. They do want to be recognized for being the dental professionals they are. Hygienists need to feel the same sense of pride and accomplishment in caring for their patients that dentists feel. Is that wrong? I don't think so.

A beautiful young woman, our associate's office manager, was accepted to hygiene school several years ago. Unexpectedly, she and her husband found out they were to be the proud parents of a little girl. She decided not to go to hygiene school at that time, but hoped to go one day in the future.

The future is here, and now this very intelligent, delightful woman, says she "would never go to hygiene school today, because of the animosity that many dentists feel toward their hygienists."

At the last dental meeting we attended, a representative of the dental society warned that "they could no longer support hygienists because several members in the local hygiene society have become militant."

How sad! Look what dentists and hygienists have done for the future of our profession. In Cosmetic Dentistry Is Our Future (Chapter 2), we proudly discuss the fact that today dentistry has become a leading profession around the world. What about tomorrow?

YOU CANNOT GIVE AWAY SOMETHING YOU DO NOT POSSESS!

- Zig Ziglar

A house divided cannot stand, at least not for very long.

Since all of the controversy arose in California and Colorado concerning independent hygiene practice, dentists in many states have been trying to take away the responsibilities their hygienists have earned.

The hygienists seem militant because their responsibilities have been threatened and in many states, denied. Many dentists are even contemplating making their dental assistants into instant hygienists!

I wonder how dentists would feel if a related profession decided that dentists were being replaced with newly "trained", less educated individuals. Dentists in states that allow denturists to practice can relate well to this situation.

IF YOU WANT TO GET THE BEST OUT OF A MAN, YOU HAVE TO LOOK FOR THE BEST THAT IS IN HIM

- Bernard Haldane

When I was accepted to hygiene school, I had been a dental assistant for several years. I naively thought I would hardly have to study because I already knew just about everything they had to teach me. You can stop laughing now.

I use this analogy as a reminder to all hygienists that believe they can practice without a dentist. We have a lot to learn from this wonderful group of professionals. In addition, by working together we can offer our patients the best that dentistry can provide. The years of education, hard work, and knowledge dentists possess have made our profession what it is today.

Since dentistry has become such a great profession around the world, we know that hygienists have contributed to some of the success. Isn't it time that dentists and hygienists acknowledge one another for our professional assets and use our combined, creative energy to move forward?

This is a book on maintaining cosmetic restorations and is certainly not intended to be political. But we hope in some way, it can warm the cold war between the dentist and hygienist and allow us to be even better and stronger, for our profession and our patients.

Lynn M. Miller, R.D.H.

Michael B. Miller, D.D.S.

INTRODUCTION

■■■◆ ◆ ◆

MAINTAINING
ESTHETIC
RESTORATIONS
PAGE xii

*I*NTRODUCTION

When I graduated from hygiene school in 1981, the last thing I ever dreamed about was writing a book for the maintenance of cosmetic dentistry. In those days, about the most unusual thing we did for the preservation of restorations was to polish amalgams with a green stone.

Upon graduating from hygiene school, I began working for a periodontist. Periodontics is an area of dentistry I truly love. As a hygienist you are able to perform at your absolute maximum every day. In addition, you have the advantage of seeing significant changes in your patients' tissue response because you see the patient two, three, or four times consecutively.

Continuing education courses are a natural part of our profession, and I naturally thought that the ones I attended taught all the newest, latest, and greatest dentistry had to offer. And they did! In the field of periodontics.

They showed me how to probe every tooth and chart a six point measurement. I marked attachment levels, furcations, bleeding, exudate, and mobility. All thorough scalings were completed with Nupro (Johnson & Johnson), carefully polishing every tooth, making sure I removed the stain. If an amalgam was rough, I would get out my little greenies and polish away. If it was still rough and worn, I would recommend the patient go to their general dentist and have it replaced.

I sincerely believed I was doing all that could possibly be done to help the patient. I certainly did not feel I was harming the patient's or the dentist's restorative work.

In October of 1986, I entered the world of cosmetic dentistry at the office of Dr. Michael B. Miller. During one of our first morning team meetings Michael tried to tell me not to use Acidulated Phosphate Fluoride (APF) or regular pumice on a particular recall patient that was coming in that day. What was this strange man talking about? I thought he possibly had a screw loose because I had never heard of this concept. Why had not my wonderful periodontist told me these things?

Michael then began telling me all of my previous hygiene polishing skills were passe' when applied to porcelain and bonding. After several arguments and being a person who does not believe anything until I see it for myself, I began doing some research. My research proved to be very insightful and successful.

I learned that I had been "helping" to shorten the life of, if not ruin, many cosmetic restorations. I decided it was time to revise the last phase of my armamentarium, polishing and fluoride and adopt some new procedures.

Even in Michael's office we took some serious measures to improve our polishing procedures. For example, we stopped using the Prophy Jet on cosmetic restorations. Study after study has proven that air-abrasive polishing instruments significantly damage bonding, porcelain, and even amalgam restorations.

After talking to my colleagues, I realized that hygienists were not being taught how to take care of cosmetic restorations. In fact, we were all ruining many of our patient's and dentist's cosmetic work.

I developed a lecture series for the hygienist, and now, two and one-half years later, I am writing a book. The strange man was right. (By the way, he was so wonderfully strange, I married him.)

A complete change in your polishing procedures will not be easy, but I assure you it will be worth every step. Polishing will become fun again because you will be able to apply different techniques and concepts, instead of the same procedure day in and day out. You will be rewarded by your patients' beautiful smiles. Your dentist and your patients will sincerely appreciate your tender loving care. Plus, this is a great practice booster!

**MAINTAINING
ESTHETIC
RESTORATIONS
PAGE xvi**

Cosmetic Dental Terminology

Cosmetic dentistry has given our profession new words, which are sometimes interchangeable. It is necessary for the student of cosmetic dentistry to understand the meanings.

The following words are given for clarification of the terminology described in the forthcoming chapters.

◆ BONDING   The physical and/or chemical adherence of one material to another. Many people think of direct resin veneers as "bonding."

◆ CERAMICS   Compounds of one or more metals with a non-metallic element, usually oxygen. They are formed of chemically and biochemically stable substances that are strong, hard, brittle, and inert conductors of thermal and electric energy. A ceramic is neither metallic nor polymeric. Ceramics are made in the laboratory under a high heating process. Porcelain and Dicor (Dentsply) are the most popular ceramics used today.

◆ COMPOSITE   A plastic material composed of inorganic filler particles (glass or silica) suspended in an organic paste. The smaller the inorganic filler particles, the higher the shine of the restoration. Composites can be chemically cured, light cured, or light activated and chemically cured (dual cure).

◆ DIRECT RESIN VENEER   Resin material is applied to the visible surfaces of an anterior of posterior tooth. The material is shaped to the optimal contour of the tooth and then is cured with a light.

◆ HYBRID COMPOSITE   Composites that have macrofilled particles mixed with microfilled particles. This results in a strong and esthetically pleasing restoration. Hybrids are used in the anterior and posterior teeth and can be polished to a reasonably high shine (but not as shiny as microfills).

◆INDIRECT RESIN VENEER   These veneers, similar in final appearance to a direct resin veneer, are made by the dental laboratory. The patient first comes in to have their teeth prepared by the dentist, just like a crown or bridge. An impression is taken, a laboratory prescription is written, and several weeks later the patient returns to have the veneers bonded to their teeth. They can be made of either a microfill or a hybrid composite.

◆MACROFILL COMPOSITE   A composite with large inorganic filler particles. Macrofills are essentially obsolete since their finish is much duller and rougher than microfills or hybrids.

◆MICROFILL   A tooth-colored composite material in which the inorganic filler particle is very finely ground silica. Microfills mimic enamel closer than any other composite, resulting in a very high shine. They are not generally recommended for high stress areas.

◆PORCELAIN VENEER   These veneers, are very similar to indirect resin veneers, but are made of a ceramic material.

◆RESIN   This term is commonly used interchangeably with "composite." Resins are actually the organic paste into which the inorganic filler particles are added.

◆SMALL PARTICLE COMPOSITE   Composites which incorporate macrofilled (large) inorganic filler particles. The two terms, macrofill and small particle, can be used interchangeably.

◆VENEER   A tooth-colored material that typically covers one or possibly two surfaces of anterior or posterior teeth. It is usually a ceramic material (porcelain or glass) or a composite resin.

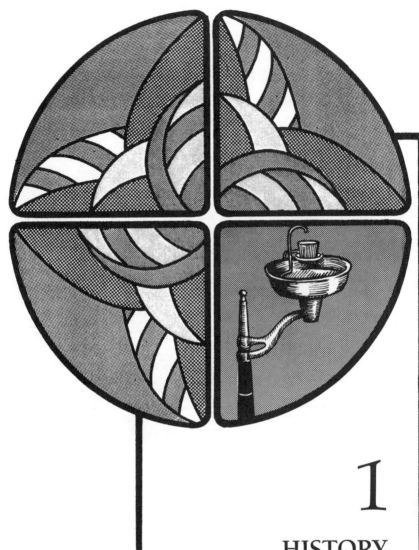

1

HISTORY
OF
COSMETIC
DENTISTRY

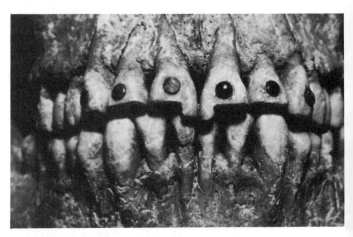

The teeth of the Mayan skull of the ninth century A.D. have numerous inlays of jade and turquoise.

Courtesy of Museo Nacional de Antropologia, Mexico City.

Chapter 1

◆ HISTORY OF COSMETIC DENTISTRY
IT'S NOT REALLY NEW

Cosmetic dentistry is really not a new phenomenon. There is evidence of cosmetic dental treatment over 15,000 years ago. Primitive societies frequently enhanced their appearance by alteration of the anterior dentition. However, they did not develop systems for the care of even the most minor dental ailments. Transformation of the dentition was usually for religious reasons, tribal identification, or esthetic effect.[1]

MAYAN DENTISTRY

In Pre-Columbian America, around 600 A.D., the Mayas were skilled in personal adornment of their upper and lower anterior teeth. Archaeologists have uncovered skulls in which beautifully carved stone inlays were placed in carefully prepared cavities in the upper and lower anterior teeth, and occasionally in the bicuspid teeth. A variety of minerals were used for these inlays, including jadette (a silicate related and similar in appearance to Oriental Jade), iron pyrites, hematite ("bloodstone"), turquoise, quartz, serpentine (when mixed with other minerals had an appearance similar to jade), and cinnabar, the ore from which mercury is extracted.[1]

We also know that the Mayas had elaborate religious ceremonies in which blackening of the teeth and sacrification of the face or torso played a part. Thus, it is reasonable to assume that tooth mutilation and ornamentation served a cultic purpose.

ROMAN DENTISTRY

As early as 450 B.C. in more developed cultures, such as the Roman Empire, a well-rounded attitude to dental care was

◆ ◆ ◆ ▬▬▬▬

HISTORY OF
COSMETIC
DENTISTRY
CHAPTER 1
PAGE 3

prevalent among upper class individuals. The Romans were proficient in the treatment of oral diseases and the extraction of teeth. They frequently restored carious teeth with gold crowns and replaced missing teeth with fixed bridgework. The Laws of the Twelve Tables suggests that prosthodontics was practiced in the early Republic of the Roman Empire. Dental restorations had become quite sophisticated and complete, as well as partial, dentures were frequently seen.[1]

Martial and Juvenal were the two principal satirists responsible for supplying the knowledge we now have of Roman dentistry. In several of Martial's writings were references to dental appliances:

> *Lucania has white teeth. Thais brown. How comes it?*
> *One has false teeth, one her own.*
> *And you, Galla, lay aside your teeth at night*
> *Just as you do your silken dress.*[1]

There is even references to practitioners who grew rich supplying artificial teeth and other prosthetic devices. It is now assumed that prosthetic appliances were fashioned by goldsmiths or other artisans, then placed in the mouth by the "physician", just as dentists and laboratory technicians share responsibilities today.[1]

BRACES

In 1757, a French dentist, Dr. Etienne Bourdet, used ivory splints to pull misaligned teeth into the correct position. He also recommended extraction of first bicuspids to alleviate overcrowding of the teeth, and then used threads attached to a splint of ivory to reposition the teeth.[1]

PORCELAIN

Two men are credited with using porcelain for the restoration of the dentition. Alexis Duchateau, a Parisian apothecary, and Dr. Nicolas Dubois de Chemant, a Parisian dentist, collaborated on their efforts to make a denture from porcelain. Finally, in 1788, Dr. Dubois de Chemant published his findings in a pamphlet form, "A Dissertation on Artificial Teeth." He then received a royal patent from Louis XVI.[1]

20TH CENTURY DENTISTRY

NONINDUSTRIALIZED "COSMETIC" DENTISTRY

Dental progress has not been made in all areas of the world. Even today, in the Amazon Valley, certain South American Indians file their anterior teeth to points to increase the ferocity of their appearance. This custom, dating back centuries, imitates the dreaded piranha fish.[1]

The Montagnards, a tribe found in the mountainous regions of Vietnam, consider normally shaped incisor teeth too "doglike". At puberty, they knock out the upper incisors or file them down even with the gum. The lower incisors are filed to a point. This custom originated as an entrance to manhood, but it is carried out today mainly for its esthetic effect.[1]

Like the ancient Mayas, natives living in the remote areas of Malaysia inlay their teeth with bits of brass wire and semiprecious stones, believing it enhances their beauty.[1]

Today, some primitive societies do appreciate and desire white teeth. The Fulani tribe of Sudan admire white, straight teeth so much they emphasize them by blackening their lips and surround their eyes with black pigment. The effect is supposedly quite startling.[1]

DENTISTRY OF MODERN CIVILIZATION

HOW BONDED VENEERS BEGAN

Most cosmetic restorations are attached to teeth by using a plastic material (usually composite resin) which interlocks into the enamel after it has been made porous by acid-etching. Two men share credit for this concept.[2]

In 1955, Dr. Michael Buonocore reported on the adhesion of plastic to enamel using acid-etching of the enamel.[2] And, in 1962, the BIS-GMA resin was introduced by Dr. Raphael Bowen.[3] It is the BIS-GMA complex that makes up many of our current composites.

LIGHT CURING

In the early seventies, light-cured composites started to replace the chemically cured materials. Light-curing gives the dentist time to place and contour an esthetic restoration before it sets. When light-cured composites were introduced, the glass particles were relatively large, causing a rough surface that shortened the esthetic life of the restoration.

Today light-cured composites have been greatly improved. The latest composites contain much smaller filler particles and have a smooth, shiny surface. Composite resins are currently being used to restore teeth in all areas of the mouth. This is done with direct restorations or indirect, laboratory fabricated ones.

HOLLYWOOD AND PORCELAIN VENEERS

While direct resin bonding has progressed at a lightning pace since 1955, porcelain has also made a big name for itself. Like bonding, porcelain has been around for a long time, but the creativity of many of today's dentists and the artistry of ceramists has made this beautiful material famous. Porcelain veneers are, arguably, the most esthetic restoration we can provide for our patients.[8]

Dr. Charles Pincus actually created the concept of placing individual thin shells of porcelain over the front teeth in 1928.[1] These were known as "Hollywood facings" and were only worn by actors and actresses during their performances, similar to makeup. One of his first patients, due to her mixed dentition, was Shirley Temple.[7]

Somewhere between 1928 and the 1980's, the concept of porcelain as a veneer withered on the vine. It wasn't until 1983 that Dr. Harold Horn reported on a technique of definitively attaching porcelain veneers to teeth. This new method consisted of etching the inside of the veneers with a strong acid. The etched veneer was then bonded to etched enamel with a resin cement.

MASTIQUE

There have been other attempts to enhance the appearance of teeth. One of the most widely used methods during the 1970's was a prefabricated acrylic veneer called the Mastique System (L.D. Caulk).[6] Developed by Dr. Frank Faunce, this procedure depended on the dentist selecting the correct size for each tooth. Mastique veneers were not custom made for each tooth and, as a result, they often did not fit properly.

Two additional problems arose with the resin used to bond the veneers to the teeth. The first problem was that the resin cement stained very badly. In addition, the resin and veneer had dissimilar adhesive properties which caused many of them to pop off. Ultimately, the Mastique veneers fell into disuse in the late 1970's.[6]

TATTOOING

Again, Hollywood was the first and last at this attempt. During the late 1970's and early 80's, tattoos were inlaid into porcelain restorations. Butterflies, flowers, initials, and various other "special orders" were attempted. This concept also withered, but like porcelain, it may one day return.[7]

POSTERIOR TEETH

The esthetic restoration of posterior teeth has also evolved into a wide variety of methods. For many years, full porcelain-fused-to-metal crowns were the only accepted tooth colored restoration for posterior teeth. In the early 80's dentists began successfully implementing new techniques for posterior teeth.

Today, all of the choices that are available for anterior teeth can be adapted for use in the posterior part of the mouth. These esthetic options include microfill and hybrid bonding, glass ionomers, and various types of ceramic restorations.

♦ ♦ ♦ ▬▬▬

HISTORY OF
COSMETIC
DENTISTRY
CHAPTER 1
PAGE 7

References:

1. Ring, M.E., Dentistry, An Illustrated History, C.V. Mosby Company, Jan. 1985

2. Goldstein, R.E., Esthetics in Dentistry, J.B. Lippincott 1976.

3. Buonocore, M.G., A simple method of increasing adhesion of acrylic filling materials to enamel surfaces, J. Dent. Res. 34:849, 1955.

4. Bowen, R.L., Dental filling materials comprising vinylsilicate product of bis-phenyl and glycolyacrylate, U.S. Patent No. 3066122, Nov. 1962

5. Horn, H.R., Porcelain Laminate Veneers Bonded to Etched Enamel, Dent. Clin. North Am., Vol. 27, No. 4, pp. 671-684, Nov. 1983.

6. Faunce, F.R., Method and apparatus for restoring badly discolored, fractured or cariously involved teeth, U.S. Patent Number 3,986,261, Filed 1973, Approved 1976.

7. Mabrito, C.A., Personal communication, April 1989.

8. Reality, Vol. 4, No.1. 1989

◆ ◆ ◆

MAINTAINING
ESTHETIC
RESTORATIONS
CHAPTER 1
PAGE 8

2

COSMETIC
DENTISTRY
IS OUR
FUTURE

CHAPTER 2

◆ COSMETIC DENTISTRY IS OUR FUTURE

A beautiful smile will be the vogue statement of tomorrow. Why? Because Cosmetic Dentistry has created a revolution! According to a recent survey done in 1987, 87% of the dentists in America alone did some type of cosmetic restoration. It is one of the fastest growing fields of dentistry. Seminars, books, and dental manufacturing companies are busy keeping up on the latest, but how to maintain beautiful smiles has not been addressed.

Thanks to fissure sealants and fluoride, carious lesions and the days of "drill and fill" are changing into the age of beautiful smiles. In the nineties, dentistry will be a word synonymous with gorgeous teeth, fitness, health, and happiness.

JOB PRESTIGE AND PERSONAL DEVELOPMENT

A recent study tested various occupational groups to determine whether they considered dental esthetics to be important for job placement and advancement. It was found there was a high relationship between job prestige and increased visibility, both of which correlated with dental appearance.[3]

Information on cosmetic dentistry appearing in the mass media translates into a more excited, astute, and demanding patient. People have a much higher dental IQ than they had ten years ago. Dental education is now easily acquired through television, magazines, and newspapers. Our patients have become more aware of the services offered to keep their health and appearance at its maximum. Cosmetic dentistry is to the teeth and lips what plastic surgery is to the entire body.

◆ ◆ ◆ ━━━━━

COSMETIC
DENTISTRY IS
OUR FUTURE
CHAPTER 2
PAGE 11

Dr. Joyce Brothers recently addressed a question from a distressed mother. The mother was upset that her daughter was not "a beauty", and, as a result had low self-esteem. Dr. Brothers says, "One facial feature we can control or change, is our smile. A beautiful smile and infectious laugh can raise self-esteem and even be sexy."[2]

This new smile phenomenon translates not only to increased patient's excitement but, subsequently, into increased production. The dental team's first responsibility lies in completely educating our patients to make them aware of which cosmetic services apply to them. After education, excitement sets in and the patient's psychological view of smile enhancement becomes open to their own smile's unique new possibilities. Next cosmetic enhancement becomes a "necessity."

There are five stages to successfully turn this psychological process into reality:

STAGE I: KNOWLEDGE

Dentistry is changing so rapidly, it is difficult for the most ardent student of the profession to keep up. Thus, it is essential that we are constantly staying abreast of the most up-to-date techniques. We then can inform our patients of the newest smile enhancement services available to them. Not all services are right for every patient: Some patients are candidates for bleaching, some for bonding, some for porcelain. Another patient might need orthodontics or periodontics before a successful cosmetic program can be completed.

We need to truly understand the interdependent role other dental professionals play in building a healthy smile. For example, if the gingival tissue is asymmetrical, is a gingivoplasty necessary before placement of veneers? Is a ridge augmentation necessary when designing a fixed bridge?

This is not just the dentist's duty. Every team member from the office manager to the dental hygienist should know these basic concepts of cosmetic dentistry. Diagnosis by team members is not being advocated, but an overall knowledge of the basic concepts so that the patient will always feel they are in a professional, well-informed office. This is the best internal marketing tool a practice can possess. It shows the patients you care and that they are not a number but an extremely important individual with very special needs.

If the Knowledge appointment is a new patient's cosmetic consultation, the patient should leave the office with a brochure concerning their specific needs. The receptionist/appointment coordinator should be available to answer questions and make the next appointment.

For established patients, the Knowledge stage is not always a "planned appointment". A team member may inconspicuously offer the information to the patient during a routine dental visit. For example, the hygienist might say, "Mrs. Jones, does the color of your front teeth bother you?" This is the art of planting a seed so it can grow into a beautiful smile. A cosmetic consultation may or may not be indicated at this time, depending on the patient's response. If it is not made, let the seed grow.

STAGE II: MOTIVATION

This is often the cosmetic consult. (Sometimes it can be Stage I and II combined. Whichever situation is applicable to the particular needs of the patient.) Patients become motivated when they begin to understand and internalize the information. During this stage, patients' objectivity ceases, and they begin to focus on the benefits of cosmetic dentistry for themselves. The dentist or team member can begin to reiterate which cosmetic enhancement treatment is applicable for the particular needs of this patient. Considerations such as the person's lifestyle, career, oral hygiene, periodontal health, and finances should be weighed in the equation before reaching a final decision.

A cosmetic consult should allow enough time to guide the patient towards a final decision. All questions must be thoroughly answered during this appointment. Educational

◆ ◆ ◆ ▬▬▬▬

COSMETIC
DENTISTRY IS
OUR FUTURE
CHAPTER 2
PAGE 13

information sheets, cosmetic brochures, and before and after pictures of patients that have received similar treatment from the dentist are all beneficial.

An appointment is made at this visit for either a prophy or the next step necessary to begin treatment. When the cosmetic consult has been completed, the patient should leave the office with a cosmetic brochure and copy of the financial arrangements. A phone call from the dentist or a staff member may be necessary to answer any questions the patient might have.

STAGE III: TREATMENT

The day of the treatment appointment is often filled with high aspirations, possibly coupled with some anxiety. It is important to ask the patient if there are any new questions since the last visit. During this appointment the dentist can be very intense so it becomes the staff's responsibility to make the patient feel comfortable and at ease. The assistant needs to convey the message, "We are here for you and when we are finished, you will look great."

It is important to give patients a sheet educating them about how to care for their cosmetic restorations. This sheet should include what to expect from their new teeth. (It often takes patients a couple of days or even weeks to get used to the new feel and look of their teeth.)

Finally, it is very important for the dentist to call the patient the same night of the treatment appointment. Ask the patient how they are feeling and how they like their new smile. This is one of the best ways to show patients that not only is the treatment #1, the dentist is too!

STAGE IV: ACCEPTANCE

The treatment is not finished when the cosmetic procedure ends. When the patient comes in for the cosmetic contouring, polishing, and the "post-operative" visit, it is important that a team member spends some time listening to the patient.

Often patients will tell the staff things they will not tell the dentist. The team member can often help the patient become psychologically aligned to their esthetic expectations,and then communicate any of the patient's questions or desires to the dentist. Satisfaction comes when we all feel understood and heard. As Omer Reed has said for years, "No one cares how much you know, until they know how much you care".

STAGE V: MAINTENANCE

It is important for hygienists and dentists to realize that cosmetic dentistry is placing a great responsibility in our hands. Maintaining the quality of dentistry our patients want and deserve is not an easy task. But it is rewarding. The rest of this book is Stage V, not only in maintaining restorations but also in constantly reaching for your limits and growing in your profession. Hopefully, it will continue the rest of your career.

Whether your niche is general dentistry, pedodontics, periodontics, or orthodontics, how to care for cosmetic restorations is now, and forever will be, a necessity.

_____

References:

1) Dental Products Report, January 1987.

2) Brothers, Joyce, The Houston Post; Are True Attractions Only Skin Deep, December 1988.

3) Jenny, J. Proshek, JM., Visibility and Prestige of Occupations and The Importance of Dental Appearance, Journal of the Canadian Dental Association, No. 12, 1986, p.987-989.

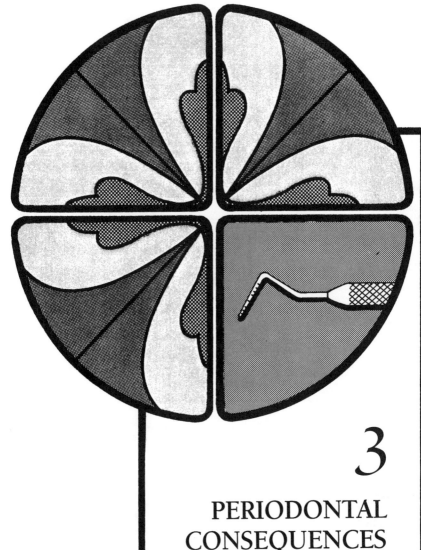

3

PERIODONTAL
CONSEQUENCES
OF
COSMETIC
DENTISTRY

■■■■■◆ ◆ ◆

MAINTAINING
ESTHETIC
RESTORATIONS
CHAPTER 3
PAGE 18

CHAPTER 3

◆ PERIODONTAL CONSEQUENCES
   OF COSMETIC DENTISTRY

Tissue adaptation to any restoration is essential in maintaining a healthy periodontium. Gingiva prefer enamel over most man-made substitutes, with the possible exception of Dicor (Dentsply). Therefore the best tissue response to restorative materials occurs when the restoration is placed supra-gingivally.[1] Since subgingival restorations are occasionally still a necessity, marginal finishing and smoothness is imperative for gingival health.

Maintaining a smooth margin around the restoration is, of course, the goal of every good clinician. To successfully maintain any cosmetic restoration, it is important to be aware of the strengths and weakness of the materials being used today as well as their idiosyncrasies in the oral cavity. Only then can periodontal health be assured.

CERAMIC RESTORATIONS, PORCELAIN VENEERS, AND GINGIVAL HEALTH

Contemporary periodontics accepts plaque as the major etiologic factor of periodontal disease. Glass ceramics and porcelain can help in maintaining gingival health as compared to other restorative materials. Restorations fabricated from these materials must have margins that fit without overhangs or marginal gaps. In addition, non-porous surfaces of glass ceramics or porcelain do not allow bacterial plaque to adhere significantly.[2]

Curettes offer the tactile sensitivity necessary to successfully clean ceramic and porcelain restorations without scratching the material. The resin cements that bond porcelain to teeth

◆ ◆ ◆ ▬▬▬▬

PERIODONTAL
CONSEQUENCES
OF COSMETIC
DENTISTRY
CHAPTER 3
PAGE 19

are not as hard as restorative resins and incorrect instrumentation can ditch the cement which leads to accelerated staining and plaque retention. Hand scalers and sonic or ultra-sonic scalers can also scratch the glaze the of glass ceramics and porcelain.[3]

Chapter 5 discusses in complete detail the rationale for treating any ceramic restoration with care. The longevity of the restoration is considered, but more importantly, tissue adaptation and health can be jeopardized when glass ceramics or porcelain veneers are cared for improperly.

DIRECT AND INDIRECT RESIN VENEERS AND GINGIVAL HEALTH

Direct and indirect resin veneers, or any composite restoration's successful tissue response depends on many variables. As with ceramic materials, the surface smoothness of composites can be altered significantly if the restorations are finished incorrectly. Other variables include the type of composite material, proper light curing, air bubbles incorporated during placement, proximity of the restoration to the soft tissue, ability to adequately isolate during bonding, and the final finishing and polishing.[4]

Gingival acceptance of the restoration depends greatly on the initial surface smoothness. Unlike glazed porcelain, tissue does not respond to composites readily. Even with a smooth surface and healthy tissue response initially, resin irregularities such as incorporated air bubbles that eventually reach the surface can cause problems later.

Direct and indirect resin veneers or any composite restoration can be softened by organic compounds found in toothpastes, plaque, food, and beverages. Softening of composites result in surface roughness and plaque retention. In addition, softening can lead to a rough margin and can cause further gingival irritation as well as excessive plaque and stain. We must impress upon our patients their responsibility in keeping composite restorations free of plaque.[5]

The maintenance program of direct or indirect resin veneers is just as important as that mentioned for porcelain veneers. This is especially true since the surface smoothness of composite resins can be more easily jeopardized by improper cleaning, polishing, and fluoride, than that of porcelain. The above warnings are not intended to paint a bleak picture.

When resin veneers are properly placed and maintained, the gingival health should be excellent. The shortcomings of the materials are being mentioned so that if surface irregularities are found during routine scaling and prophylaxis, the dentist can be alerted to these findings.

Chapter 6 discusses in complete detail the various composites and their different strengths and weakness. Chapters 8 and 10 will answer various other questions concerning the required maintenance for composites.

As previously discussed in the "Glass Ceramics, Porcelain Veneers, and Gingival Health" section, curettes are also the hand instrument of choice for all resin bonded veneers and composite restorations.

STEPS TO SUCCESS FOR PERIODONTAL HEALTH

APPOINTMENT #1:

A thorough scaling, root planing, and periodontal probing is necessary before the patient begins treatment for any esthetic restoration. Unhealthy tissue will not respond well to the manipulation necessary for preparation of the restoration.[6]

Periodontal Charting

It is important to chart all periodontal pockets (even in the 3mm range), attachment levels, recession, and inadequate attached gingiva. Chart all areas where bleeding and exudate are found, particularly in the area where the cosmetic restorations are to be placed. This will be a good protection from any medico-legal problems in the future.

◆ ◆ ◆ ▬▬▬▬

PERIODONTAL
CONSEQUENCES
OF COSMETIC
DENTISTRY
CHAPTER 3
PAGE 21

If any pockets or inadequate attachment levels are found, it might be necessary to consider periodontal treatment for the patient before cosmetic treatment begins.

Dentist/Hygienist/Patient Communication

Your dentist is relying on your judgement and feedback. Document your periodontal findings by written information in the patient's chart. Document all discussions with the patient including areas of periodontal concern as well as oral hygiene instructions. Poor oral hygiene often indicates that direct or indirect resin veneers would not be a good choice for the particular patient. In addition, discuss any concerns you might have about your patient's periodontal health with your dentist.

Oral Hygiene Instructions and Plaque Indices

Oral hygiene skills need to be at their maximum before recommending cosmetic restorations. Chart plaque indices and document all oral hygiene instructions in the patient's chart. An ultra-soft toothbrush is recommended for all cosmetic restorations to minimize scratching or abrasion. Flossing becomes even more important, due to the harmful effects of plaque and acid on the proximal, surfaces of composites and porcelain. The use of a Rota-dent (Prodentec) is also advised. The Rota-dent's small soft brushes allow access to all tooth surfaces, including subgingival, ensuring a virtually plaque-free environment with stimulation to the tissue.[8]

♦ ♦ ♦

Insurance Tip: Many insurance companies cover some or all of the cost of a Rota-dent. File as a 4360.[7]

♦ ♦ ♦

Re-evaluation Appointment

Patients with gingivitis will need a re-evaluation appointment to insure the edema and bleeding have subsided. If the tissue is healthy at the re-evaluation appointment, cosmetic procedures can begin. Reiterate the need for excellent oral hygiene for the maximum longevity of the restorations. (A re-evaluation appointment is also necessary if the patient has poor hygiene habits, especially when any type of cosmetic restoration is being considered.)

If the gingivitis is persistent, check for overhanging margins on already placed restorations, plaque level, and oral hygiene skills. Metabolic disruptions such as medication, viruses, diabetes, and hormonal changes can all contribute to gingivitis. Ultimately, the patient's gingival health and oral hygiene skills need to be at their maximum before cosmetic preparation begins.

APPOINTMENT #2: PREPARATION OF THE COSMETIC RESTORATION

The cosmetic preparation appointment, for resin or porcelain restorations, should be about a week to ten days following the last cleaning to allow for tissue healing.

If, for some reason, the patient cannot make the second appointment until four to eight weeks after the last cleaning, a re-evaluation appointment may be necessary to ensure the tissue is healthy. If the gingiva was bleeding with edema or exudate, the re-evaluation appointment is mandatory.

Recently our office had to remove eight porcelain veneers on a young woman (See Fig. 1) because of the severe gingivitis they were causing. An initial scaling and root planing had not been performed. The dentist who placed the veneers told her, "these sometimes cause gingivitis!" And they did.

◆ ◆ ◆

VENEERS SHOULD NEVER CAUSE GINGIVITIS. IF GINGIVITIS OCCURS AROUND A VENEER, CHECK, AND IF NECESSARY, CONTOUR ALL MARGINS AND PERFORM A PERIODONTAL SCALING.

◆ ◆ ◆

◆ ◆ ◆ ▬▬▬▬

PERIODONTAL
CONSEQUENCES
OF COSMETIC
DENTISTRY
CHAPTER 3
PAGE 23

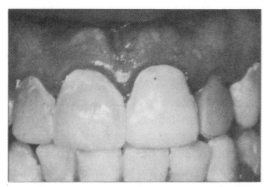

FIGURE 1: Patient presented edema and bleeding after placement of first porcelain veneers.

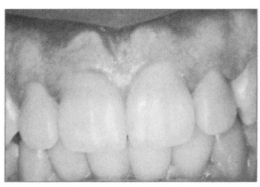

FIGURE 2: Healthy tissue is now present: two weeks after removing veneers and performing two periodontal cleanings.

◆ ◆ ◆

MAINTAINING
ESTHETIC
RESTORATIONS
CHAPTER 3
PAGE 24

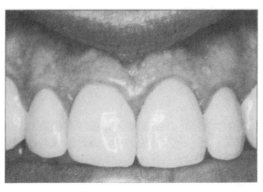

FIGURE 3: New porcelain veneers. Excellent contouring of gingival margins, as well as future regular cleanings, will provide maximum health for the tissue.

After the veneers were removed, two thorough scalings were performed (See Fig. 2). After a week of healing her teeth were prepared for new veneers. She now has beautiful porcelain veneers surrounded by healthy pink stippled tissue (See Fig 3).

Once cosmetic restorations are placed, good oral hygiene and three to four month recalls are a necessity to maximize longevity of the *esthetics* of the restoration. Remember, the esthetic life of a restoration and the functioning life can be quite different. If the maintenance program is not carried out properly, the esthetic life can compromised.

ANTIMICROBIAL RINSES

Antimicrobial rinses have become very popular for gingivitis and more advanced periodontal disease. They can indeed be very useful in helping to eliminate, or at least reduce, anaerobic bacteria in the oral cavity.

Most of these rinses contain a very high amount of alcohol. Alcohol softens bonding which can lead to further plaque retention, staining, and microleakage.[8]

Peridex (Proctor and Gamble), which is .12% chlorhexidine gluconate, not only contains a high amount of alcohol, but can cause staining of the teeth. This staining is due to its attachment to the tooth, replacing the pellicle and micro-organisms. Peridex is, however, an excellent rinse to combat gingivitis and reduce inflammation. Between the first scaling and root planing and re-evaluation appointment (before preparation or placement of any cosmetic restoration), Peridex can be used as a home rinse, if necessary.[9]

Unfortunately, the residue of Peridex adheres readily to composites, due to their inherent porosity. Therefore, after placement of porcelain veneers (which have resin cement around them) and resin restorations, refrain from having patients rinse with Peridex at home. If Peridex needs to be used as an irrigant for specific sites, then the tooth involved should be

◆ ◆ ◆ ▬▬▬▬

PERIODONTAL
CONSEQUENCES
OF COSMETIC
DENTISTRY
CHAPTER 3
PAGE 25

polished thoroughly immediately following the procedure. A safe home rinse for the patient with periodontal problems and resin restorations or porcelain veneers is hydrogen peroxide.

Patients that have specific sites with deep pockets will benefit from a home irrigation device. This will allow patients to use the irrigant of your choice, without damaging the cosmetic restoration.

◆ ◆ ◆

Insurance Tip: If there are any pocket depths greater than 3mm, with bleeding and edema, one of the extra cleanings could be filed as a 4341 for the quadrant with the pockets.

◆ ◆ ◆

The Issue is, First the Tissue.

References:

1. Orkin, D.A.; Reddy,J.; Bradshaw,D., The relationship of the position of crown margins to gingival health. Journal of Prosthetic Dentistry, Vol.57, No.57, April 1987,pp.421-424.

2. Malament, K.A., Considerations in Posterior Glass-Ceramic Restorations, The International Journal of Periodontics and Restorative Dentistry 4/88, pp. 33-49.

3. Vitasek, B.; Acoria, L.: Ferracane, J., Profilometric Evaluation of Resin Surfaces following Hand Instrumentation, I.A.D.R. Abst. #866, March, 1988

4. Eide, R.; Rveit, A.B., Finishing and polishing of composites. Acta Odontol Scand 1988; 46:307-312.

5. Waghalter, R., The Perio Probe, August 1988.

6. Van Diken, J.W.V.; Sjostrum S.; Wing K., The effects of different types of composite resin restorations on marginal

■■■■◆ ◆ ◆
MAINTAINING
ESTHETIC
RESTORATIONS
CHAPTER 3
PAGE 26

gingiva, Journal of Clinical Periodontology 1987; 14; 185-189.

7. Van Der Linden E.; Cross-Poline G.; Tilliss T.S.; Stach D.; Featherstone, M.D., The Efficacy of the Rota-dent Compared to a Conventional Toothbrush, I.A.D.R. Abst. #2289, March 1, 1988.

8. Asmussen, E., Softening of BIS-GMA based polymers by ethanol and by organic acids of plaque, Scandinavian Journal of Dental Research 1984; 92:257-61.

9. Greenstein, G.; Berman C.; Jaffin R., Chlorhexidine, An Adjunct to Periodontal Therapy, Journal of Periodontology, Vol.57,No.6, June 1986,pp.370-377.

◆ ◆ ◆ ▬▬▬▬▬

PERIODONTAL
CONSEQUENCES
OF COSMETIC
DENTISTRY
CHAPTER 3
PAGE 27

4

GENERAL
PROTOCOL
AND
PROCEDURES

CHAPTER 4

◆ GENERAL PROTOCOL AND PROCEDURES

Successfully implementing the maintenance of cosmetic procedures will involve four basic areas of the dental office:

I. EDUCATING THE PATIENT

II. EDUCATING THE STAFF

III. PREPARING THE PATIENT'S CHART

IV. SETTING UP THE OPERATORIES

I. EDUCATING THE PATIENT

Before placement of any cosmetic restoration, our patients need to be completely informed of exactly what each appointment will entail. This educational process should include all aspects of oral hygiene, length of treatments, what to expect after each treatment, and the long-term commitment to maintaining the restorations.

1. Tell patients how their teeth will "feel" after indirect or direct resin veneers, resin restorations, diastema closures, or porcelain veneers. For example, teeth that are prepared for porcelain veneers feel rough, thin, and "fuzzy". Temporaries usually do not look as good as the "real thing." New resin veneers often feel thick and look "big." These awkward feelings go away very quickly after the placement appointment.

If the patient is told ahead of time, they will be mentally prepared for the expected outcome. If they are not

◆ ◆ ◆ ▬▬▬▬▬
GENERAL
PROTOCOL AND
PROCEDURES
CHAPTER 4
PAGE 31

informed, many patients feel something is wrong or the dentist did not do a good job.

2. Patients need to have their teeth cleaned three to four times a year (depending on their level of hygiene ability and peridontal health) after placement of resin or porcelain veneers.

3. Patients need to be aware of the varying amounts of time resin or porcelain veneers will last. With proper maintenance, it will be much longer than if they do not care for them. (This decision is best left to the individual dentist. Each dentist has their own comfort level as to the time recommended.)

4. Patients need to know that the dental team has been specially trained to keep their new restorations beautiful. Don't be afraid to brag a little bit!

5. After the cosmetic procedure, give all patients a "CARING FOR YOUR COSMETIC RESTORATIONS" sheet.

Have your patient sign a form stating that they have been informed of the necessary information before cosmetic treatment begins. Give a copy to your patient and keep one in the chart.

II. EDUCATING THE ENTIRE TEAM

In addition to the hygienist, it is essential that the rest of the dental team become aware of the new procedures being implemented in the office.

◆ ◆ ◆

MAINTAINING
ESTHETIC
RESTORATIONS
CHAPTER 4
PAGE 32

A. Appointment Coordinator/Receptionist

1. The appointment coordinator should know how to properly communicate to and schedule patients for cosmetic procedures. The coordination and timing of the necessary procedures usually begins at the front desk. For example, the initial cleaning should be done 7 to 10 days before the cosmetic placement/preparation appointment. If a patient wants to hurry up the treatment and step out

of sequence with the doctor's treatment plan, a well educated appointment secretary will be able to communicate the benefits of following the proper sequence. An educated staff saves your office and patients time, trouble, and money!.

2. Teach every member of the team how to answer the telephone in a professional manner.* (Cross-training of telephone techniques is very important to a successful cosmetic practice.) Make sure the answers your patients are getting are the answers you want them to hear. For example, if a new patient calls and asks, "Does your office know how to take care of my veneers?", will get the correct answer.

3. The office manager should be instructed in cosmetic recall procedures. Three months after the placement appointment is when the patient needs to come in for their first maintenance recall appointment. The office manager should understand the reasons to schedule this appointment and gently remind patients that it is time for their cleaning, if they have forgotten.

* Strawberry Communications offers excellent tapes including how to properly answer the telephone, internal and external marketing on the phone, how to handle an irrate caller, and much more. You can order these tapes through Zig Ziglar, Dallas, Texas.

B. Dental Assistant

Dental Assistants need to know the correct protocol for polishing porcelain and resin restorations. (In many states, dental assistants are allowed to polish coronal tooth structures and restorations.) Below is a list of essentials for polishing cosmetic restorations, whether the assistant, the doctor, or the hygienist implements them.

◆ ◆ ◆ ▬▬▬▬

GENERAL
PROTOCOL AND
PROCEDURES
CHAPTER 4
PAGE 33

FOR DIRECT AND INDIRECT RESIN RESTORATIONS

1. Put one rubber cup (polisher) in the latch-style handpiece. Have one rubber instrument cup (finisher) on the tray, with two rubber points (one finisher and one polisher).

2. Dispense a small amount of aluminum oxide polishing paste for the composite veneer. Dispense one small amount of prophy (pumice) polishing paste for enamel.

3. Put one prophy cup (latch-style) and one prophy brush (latch-style) on the tray. Dispense one 10-inch piece of unwaxed floss.

4. Dispense one application of sodium fluoride.

FOR PORCELAIN VENEERS

1. Put one rubber cup (polisher) in the latch-style handpiece. Have one rubber instrument cup (finisher) on the tray, with two rubber points (one finisher and one polisher).

2. Dispense a small amount of diamond polishing paste for the porcelain veneer. Dispense a small amount of prophy (pumice) paste for enamel.

3. Repeat steps 3 and 4 above.

III. PREPARING THE PATIENT'S CHART

A. Apply the "Porcelain" or "Bonding" sticker to the front of the patient's chart. We designed this to be used like a "medical alert" so that the doctor, the staff, or even a temporary hygienist could become instantly familiar with this patient and their restorations' special polishing needs. (It also speeds up setting up the operatory.)

B. It is now time to incorporate the three to four month cleaning regimen. Mark it in the chart.

IV. SETTING UP THE OPERATORIES

Operatories that will be used for polishing bonding or porcelain should have the following:

A. BONDING or PORCELAIN alert sticker (Reality)

B. Kits: Esthetic Maintenance Kit (Shofu)

C. Latch-type slow speed handpiece

D. Prophy cups

E. "Caring for Your Cosmetic Restorations" Sheet (Reality)

F. Composite Polishing Paste

Composite polishing paste is aluminum oxide particles incorporated in a semi-viscous paste

1. Prisma - Gloss (L.D. Caulk)
2. Composi-Glaze (Denpac)
3. Command Ultra-Fine Luster Paste (Kerr)
4. Enamelize (Cosmedent)

G. Porcelain Polishing Paste

Porcelain polishing paste contains very finely ground diamond particles in a semi-viscous paste.

1. Save-a-Bake Diamond and Porcelain Polish (Denpac)
2. Truluster Polishing System (Brasseler)
3. Diaglaze (Antraco)
4. Porcelain Laminate Polishing Paste (Den-mat)

H. Sodium Fluoride

I. Curettes

◆ ◆ ◆ ▬▬▬

GENERAL
PROTOCOL AND
PROCEDURES
CHAPTER 4
PAGE 35

J. Plastic Instrument

K. Rubber Polishing Instruments

COMPOSITE RUBBER POLISHING INSTRUMENTS

1. Mini-Identoflex (Centrix)
2. Polishing Points, Cups, Wheels (Kulzer)
3. Polishers and Finishers (Vivadent)
4. Quasite Midi-Point and CompoSite Points (Shofu)

PORCELAIN RUBBER POLISHING INSTRUMENTS

1. Ceramiste Silicone Points (Shofu)

L. Polishing Strips (Choices)

1. Flexistrip (Cosmedent)
2. Sof-Lex Finishing and Polishing Strips (3M)
3. Epitex (I.C.E./COE)

M. Polishing Discs

1. Sof-Lex XT Pop-On Extra Thin Contouring and Polishing Discs (3M)
2. FlexiDisc Mini Discs (Cosmedent)
3. FlexiDisc All Purpose Finishing and Polishing System (Cosmedent)
4. Sof-Lex Pop-On Contouring and Polishing Discs (3M)
5. Super-Snap (Shofu)

N. Magnification Lenses

1. Zeiss (Den-Mat)
2. Surgical Binoculars (Designs for Vision)
3. Orascoptic Telescopes (Advanced Dental Concepts)

◆ ◆ ◆

MAINTAINING
ESTHETIC
RESTORATIONS
CHAPTER 4
PAGE 36

Protocol and Procedures Can be Boring Because They are Inanimate. What Will Bring Them to Life? You.

GET STARTED AND START SMILING!

5

THOSE
PRETTY
PORCELAIN
VENEERS

CHAPTER 5

◆ THOSE PRETTY PORCELAIN VENEERS (AND CROWNS, INLAYS, ETC.)

VENEERS

Porcelain Veneers offer our patients, arguably, the most esthetic, trouble-free and longest-lasting conservative restoration to enhance the appearance of their teeth. Since they are baked in an oven at high temperatures, porcelain veneers have a very hard surface that resists staining. Porcelain is usually colored from within which means there is very little chance of removing the color during polishing.

PREPARATION AND CEMENTATION

Preparation of the tooth is usually necessary for the veneer to maintain gingival health and a natural appearance. A pre-operative cleaning is necessary one week to ten days before his appointment to ensure the gingival health is at its maximum, free of plaque or calculus. On the day of the preparation appointment, the teeth should again be cleaned with pumice.

Typically, for the gingival portion of the tooth, .3-.5 mm of enamel is removed while .5-.7 mm of enamel is removed from the incisal portion of the tooth. Reduction depends on the pre-operative condition of the tooth. Facially inclined teeth need to be prepared to a greater extent than lingually inclined teeth. After the preparation, an impression is taken of the prepared teeth and the opposing teeth. (Laboratories depend on a well prepared tooth, a good impression, and a detailed written prescription to give the dentist a veneer that fits perfectly.) Temporary veneers are usually not necessary since a thin layer of enamel is left for protection. Some veneer preparations do expose dentin. When dentin is exposed, patients may experience varying levels of sensitivity.

◆ ◆ ◆ ▬▬▬▬

THOSE PRETTY
PORCELAIN
VENEERS
CHAPTER 5
PAGE 39

Cementation

The resin cement used in bonding the veneer is often the "key to success" for the final color. When the correct shade or shades are chosen to be used under the veneer, the result can be beautiful. However, a porcelain veneer can return from the lab with the correct shade, only to be ruined by poor shade selection of the resin cement. Thus the final result is a combined effort: the coloring in the porcelain and the coloring in the resin cement. In order to obtain the desired effect, it is often necessary for the dentist to select two resin cements to be mixed together. For example, many cosmetic dentists chose a translucent shade and an opaque shade, or a whiter shade with a yellow shade, depending on the patient's needs and desires.

The resin cements used for bonding porcelain veneers are more fluid versions of composites, most often a hybrid composite. These resin cements are either totally light cured or dual cured. Light cured composites have less porosity since there is no mixing involved. The dual cured composites have to be mixed and therefore can incorporate air during mixing which can lead to increased porosity. However, to obtain the correct shade, different colors of light cured resin cements often are mixed together. This increases the risk of additional porosity. (See "POROSITY" Chapter 6)

Contouring

Optimally, all veneers should fit perfectly and not need adjustments. But, in reality, many veneers will need:

◆ ◆ ◆

MAINTAINING
ESTHETIC
RESTORATIONS
CHAPTER 5
PAGE 40

1. Minor adjustments to the incisal edge and line angles to help customize the veneers to the patient's face.

2. Finishing at the gingival margins to ensure optimal gingival health.

Shaping a porcelain veneer in the mouth is not easy and it is impossible to re-glaze it once it is bonded to the tooth. Adjustment of a bulky gingival margin or emergence profile is done with a finishing diamond followed by a 30 fluted carbide finishing bur. (The flutes in a bur are the cutting edges, the higher the number, the smoother the bur. Normal operative burs have six flutes.)

It is important for the operator to note that these adjusted areas are the most susceptible to stain. This is especially true of the gingival margin. Not only does it accumulate plaque and other debris at a higher rate than other parts of the tooth, it can also have the added insult of a margin composed of excess resin cement. (If the laboratory finished the gingival margin improperly or the final fit was just not adequate, resin cement will fill the gap between the veneer and the tooth.)

Because they are less filled than regular composites, the resin cements used to bond veneers are not as resistant to staining and tooth brushing as regular composites. Therefore, because of abrasion and wear of the resin, a gap may eventually form at the interface between the veneer and the tooth. This gap typically stains at a faster rate than the remainder of the veneer. The gingival margin is thus considered the weak link of the veneer.

There may also be a gap between the veneer and tooth at the proximal line angles. Stain may lodge in these areas as well. The untouched portion of the veneer which still retains its glaze is the least likely area to stain. However, small microcracks can be present in veneers and can stain.

CROWNS

Traditional porcelain crowns are made in the laboratory by baking porcelain over a core of metal. This design gives the crown tremendous strength and typically fits very well. These crowns are most often cemented with a more conventional type of cement. Even though these cements can also be abraded, there is usually less cement exposed due to the superior fit of cast metal (under the porcelain) to the tooth.

However, the new all-ceramic (porcelain) crowns have no metal substructure. Since the fit of these all-porcelain crowns

◆ ◆ ◆ ▬▬▬▬

THOSE PRETTY
PORCELAIN
VENEERS
CHAPTER 5
PAGE 41

usually does not quite match the fit of a precise metal coping, the gingival margin may have more exposed resin cement. Like porcelain veneers, this resin cement at the margins has a propensity to stain or attract plaque.

INLAYS AND ONLAYS

These restorations are all-ceramic and are etched on the inside. They are bonded to teeth just like veneers and must be finished in the mouth. This means finishing not only their margins, but their occlusion. Unless they fit extremely well, the exposed resin cement at the margin will wear at a higher rate than the porcelain. This means that there can be gaps at all margins, including the occlusal.

One very difficult area for a dentist to finish properly is the gingival margin. If there is an overhang of porcelain that was not found before bonding, it should be removed after bonding. This procedure is very tricky and can result in ditching of the tooth and/or the restoration. At the very least, it will remove the original glaze, which can never be replaced.

If the tooth and the restoration are ditched, it is almost impossible to smooth the area without a special handpiece (Profin, Weissman Technology). Even with this handpiece, the results are less than perfect. The "smoothed" areas of the inlay or onlays, now unglazed, have a much higher tendency to stain or attract plaque.

CERAMICS WITH A UNIQUE COLORING SYSTEM

DICOR

Dicor is a relatively new cast ceramic. Gingival response to Dicor is unmatched, so it is "heaven" for the hygienist. Since Dicor is the only ceramic that is totally shaded on the surface, it must be carefully polished. Continuous use of coarse prophy pastes or toothpastes could possibly remove the color of Dicor.

ROUTINE SCALING AND POLISHING OF PORCELAIN OR DICOR

◆ ◆ ◆

Clinical Tip: Use at least 2-power magnification for routine scalings and polishings. The difference is amazing!

◆ ◆ ◆

Porcelain and resin cements can develop tiny cracks. In addition, these restorations are frequently used to close diastemas or create other optical illusions. Magnification helps detect the differences in the surfaces.

Step 1: Curette carefully around a veneer. A curette's tactile sensitivity optimizes esthetics by leaving a smooth restoration, free of plaque and calculus. Scalers should not be used on porcelain because they can scratch the glaze, hook a margin, or abrade and ditch the resin cement.

Step 2: Apply a diamond polishing paste directly to the porcelain. Use enough to thoroughly cover the restoration, but don't waste it - it is expensive.

Step 3: Polish each restoration thoroughly for 30 seconds, using a rubber cup.

Step 4: With the paste still on the restoration, floss the teeth to carry the paste interproximally.

Step 5: Wash the paste, dry the teeth, and inspect the restorations. You are finished if there is no visible stain.

Step 6: Have the patient rinse for four minutes with sodium fluoride.[1]

WHEN NOT TO USE DIAMOND POLISHING PASTE: If the gingival margin has a noticeable gap filled with resin cement, use composite (aluminum oxide) polishing paste for

◆ ◆ ◆ ▬▬▬

THOSE PRETTY
PORCELAIN
VENEERS
CHAPTER 5
PAGE 43

the entire tooth. Composite polishing paste will not harm porcelain. Diamond polishing paste will harm resin cements.

If there is stain after polishing with diamond paste:

Proximal Stain:

Light and heavy proximal stain or roughness can be removed with polishing strips. Start with the finest grade. If this does not remove the stain, graduate to the next coarsest grade. Continue this process until the stain is removed. Then graduate back up to the finest grade to leave the restoration as smooth as possible. The coarse grade of the strips should not be used for stain removal as they can open an embrasure. Carefully insert the non-abrasive middle portion of the strips into the embrasure so a contact is not opened. (Epitex Strips do not have a non-abrasive middle, but are very thin and easy to insert.)

Heavy and Medium Stain:

Besides proximal areas, heavy stain in all other areas can be removed with polishing discs, including embrasures. These discs come in two sizes and four grits: coarse, medium, fine, and extra-fine. (The coarse are not recommended for this procedure.) Discs should be used only after the operator has been carefully trained so the glaze of the porcelain will not be removed or ditching of the resin or porcelain will not occur.

1. Start with the medium grit. For gingival areas, protect the tissue with a plastic instrument. Apply light even pressure to the stained and roughened surface for 10 to 15 seconds. Rinse and carefully examine the tooth. You are ready for Step 2 if there is no visible stain or feeling of roughness.

2. Use the fine grit for the same time the medium grit was used. It is important to cover the same surface area of the tooth to ensure a smooth surface.

3. The extra-fine is now used for the same amount of time the fine and medium were used. Thoroughly rinse the tooth for 30 seconds to ensure all residue is gone.

4. Polish the surface areas treated with diamond polishing paste. Polish for 30 seconds, carrying the polish interproximally with floss, if needed.

Light Stain

Porcelain Rubber Polishing Instruments are slightly harder than composite rubber polishing instruments. Ceramisté by Shofu are used in the example below. They come in three grades, but the two finest should be used for this procedure.

1. For facial or subgingival surfaces start with the Ceramisté Ultra Cup. With the tooth dry, use light, intermittent pressure until the stain is gone. Carefully examine the tooth visually and with an explorer, to ensure there is no visible stain or feeling of roughness.

2. Use the Ultra II Ceramisté Cup, carefully polishing the same tooth areas covered with the Ultra Cup.

3. For lingual, incisal, or occlusal surfaces start with the Ceramisté Ultra Midi-points. Begin with the Ultra and graduate up to the Ultra II. Polish with each until the stain disappears.

◆ ◆ ◆ ▬▬▬▬▬

Always use light, uniform strokes with rubber polishing instruments. Too much pressure can overheat the restoration and the tooth.

Apply the diamond polishing paste to all previously instrumented tooth surfaces. Polish thoroughly for 30 seconds with a rubber cup. If strips or discs went interproximally, carry the paste in those areas with floss.

Dry the tooth and carefully inspect the restoration. You are finished if there is no visible stain.

PORCELAIN AND CERAMIC NO-NO'S

1. Never use coarse polishing pastes. They can scratch and dull the glaze of the porcelain and abrade the resin cement.

2. Never use a sonic (example: Titan, SS White) or ultrasonic scaler (example: Cavitron, Dentsply). They will scratch the veneer and remove significant amounts of the resin cement. This can lead to breaking the bond of the veneer attachment to the tooth.[2]

3. Never use acidulated phosphate fluoride as it can etch the veneer and resin cement. This fluoride has been found to increase surface roughness, as well as lead to loss of luster. The result will accelerate pitting and staining.[3,4]

4. Never use an air abrasive polishing system (example: Prophy-Jet, Dentsply). These can break the resin bond and harm the glaze, causing pitting, staining, and loss of luster.[5]

5. Never try to remove a rough edge or "overhang" on a veneer. Remember, veneers are often placed to close diastemas and/or straighten a previously mal-positioned tooth. Always check with the dentist if there is a question about an overhang, deep stain, crack, or rough edge.

Miscellaneous

Porcelain techniques are becoming more and more advanced. Dentists have learned to refine etched porcelain so that partial instead of full veneers can be bonded to teeth.

We cannot mention every situation that you would encounter, but two simple rules should be applied:

1. Diamond polishing paste should be used when porcelain and enamel are together. Treat the entire tooth as a porcelain veneer.

2. Composite (aluminum oxide) polishing paste should be used when porcelain and composite are placed together on a tooth. Diamond polishing paste will roughen

composites, but aluminum oxide paste will not harm porcelain.

Chipped porcelain veneers can be repaired with small segments of porcelain made from an impression of the prepared fracture or with composite directly in the mouth. The care of any repair would not be any different than a regular veneer. However, repairs, whether composite or porcelain, are not as strong as the original restoration.

Damage to Porcelain by Rubber Dam Clamps

Bleaching of teeth has become a very popular and successful procedure to lighten and brighten teeth. Frequently when we bleach bicuspid to bicuspid, it becomes necessary to put a rubber dam clamp on a bicuspid or molar that might have a porcelain fused to metal crown (PFM). Studies have shown that a rubber dam clamp can damage porcelain, even though the damage may not be readily apparent. When this situation occurs, it is advisable to clamp adjacent teeth (assuming they do not have crowns) or use floss ligation.[6]

Efficiency is Doing Things Right.
Effectiveness is Doing the Right Things.

References:

1) Ripa, Louis, Effect of Type of Fluoride Compound and Fluoride Concentration on the Caries Inhibition of Dentifrices, Compendium of Continuing Education in Dentistry, Supplement No. 11, 1988, pp. 365-370.

2) Zitterbart, Paul, Effectiveness of ultrasonic scalers: A literature review. General Dentistry, July-August 1987, p. 295-297.

3) Gonzalez, E.; Naleway; C.A., Fan; P.L.; Jaselskis ,Decrease in reflectance of porcelains treated with APF gels. Dental Materials 1988: 4: 289-295

(4) Council on Dental Materials, Instruments and Equipment; Status report: effect of acidulated phosphate fluoride on porcelain and composite restorations; JADA, Vol. 116, Jan.1988, p. 115.

(5) Barnes, C.M., Hayes, E.F., Leinfelder, K.F., Effects of an air-abrasive polishing system on restored surfaces, General Dentistry/May-June 1987, pp. 186-189.

(6) Madison, S.; Jordan,R.D.; Krell, K.V., The Effects of Rubber Dam Retainers on Porcelain Fused-to-Metal Restorations. Journal of Endodontics, Vol.12, No. 5, May 1986 pp.183-186.

■■■■■■■ ◆ ◆ ◆

MAINTAINING
ESTHETIC
RESTORATIONS
CHAPTER 5
PAGE 48

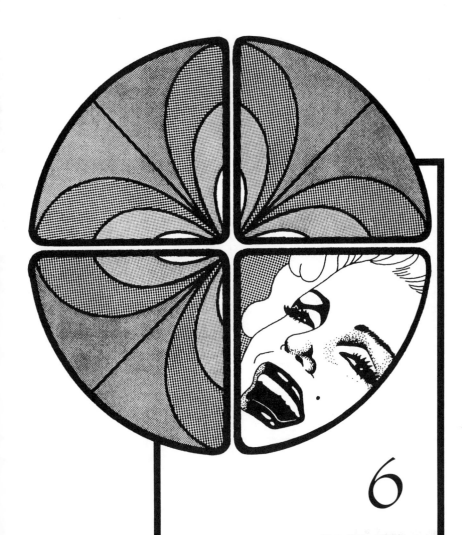

6

MARILYN MONROE
OF PLASTICS
AND
THE SUPERMAN
OF PLASTICS

■■■◆ ◆ ◆

MAINTAINING
ESTHETIC
RESTORATIONS
CHAPTER 6
PAGE 50

CHAPTER 6

♦ THE MARILYN MONROE OF
  PLASTICS: MICROFILL COMPOSITES
  THE SUPERMAN OF PLASTICS:
  HYBRID COMPOSITES

Bonding

Bonding is the adhesion of one material to another such as that of a porcelain or resin veneer to enamel. However, the putty-like material used to form posterior and anterior restorations, including resin veneers, is more commonly thought of as "bonding". This material is technically called a composite resin or just "composite". The bond of composite resins to enamel is extremely strong.

Composites are inorganic filler particles (some type of glass) mixed with organic resin to form a paste. The inorganic filler particles come in many sizes and shapes. The finer the filler particles, the higher the shine of the final restoration.

Porosity

The mixture of the inorganic and organic materials cause a certain level of porosity, or entrapment of air. When a composite restoration is porous on the surface, plaque and stain retention are greatly increased. Should air become trapped inside the restoration, breakage or microleakage can occur.

The porosity problem of composites has been greatly improved in the past several years by both researchers and dental manufacturers. Studies have shown smaller filler particles cause less porosity. Dental manufacturers are now vacuum mixing composites to try to eliminate air entrapment.

♦ ♦ ♦ ▬▬▬

MICROFILL AND
HYBRID
COMPOSITES
CHAPTER 6
PAGE 51

Even with the vacuum mixing and the finer filler particles, porosity remains a problem with composites. Additional porosity can occur with improper placement technique of the restoration.

LIGHT CURING

Light curing has been one of the main factors in advancing the cosmetic era of dentistry. In the " old days", composites were chemically cured which limited the working time. Also the chemicals added to this generation of composites, which were necessary for the composite to harden, caused premature yellowing and darkening.

Today, most composites are light cured, which means they harden or cure only after a curing light is placed on them. This allows for a much longer working time for the dentist. Enhanced esthetics and less tendency for the restoration to yellow or darken result from this process. Light curing also eliminates mixing which is a major contributor to air entrapment in the cured composite. This air entrapment can then lead to voids showing up on the surface of the restoration. Voids on the restoration's surface will increase its staining potential.

Microleakage

A small void can occur between the tooth and the restoration resulting in microleakage. This microleakage can result in increased tendency for further decay if bacteria are allowed to penetrate this void.

◆ ◆ ◆

MAINTAINING
ESTHETIC
RESTORATIONS
CHAPTER 6
PAGE 52

Microleakage can be a problem with many composites. This can be due to the inherent porosity, placement techniques, polymerization shrinkage, conditions in the oral cavity, and maintenance by the patient and the dental office.

For the hygienist, it is important to be able to recognize microleakage versus stain. The two can look similar. Microleakage is found at margins of composite restorations especially where enamel is absent, such as root surfaces. It can spread underneath the restoration giving the tooth and the composite a dark appearance. Microleakage can be asymptomatic or can cause sensitivity to the patient when eating, drinking, and inhaling.

Weak Areas

The gingival margin is naturally the area most often curetted or root planed during a routine prophylaxis. This is the weakest portion of any composite resin restoration. As the composite resin is contoured, finished and polished, the gingival margin is thinned out and can be susceptible to breakage. This part of the composite restoration is known as the weak link of the restorationn.

THE MARILYN MONROE OF PLASTICS: MICROFILL COMPOSITES

Microfill composites have been the primary force that has revolutionized dentistry from a science to becoming truly an art. Microfills are the materials most often used for direct and indirect resin veneers. Many dentists prefer direct resin veneers over porcelain veneers because the control of the restoration is kept in the dentist's hands, not in the laboratory.

Microfill composite resins have a filler particle size of 0.04 microns and are the most polishable of all the resins. In general, they sacrifice strength and durability for polishability. At least one manufacturer, however, has produced a microfill that is reinforced and may actually be as strong as some hybrids.) Microfills were originally designed strictly for esthetics and are still most often used for anterior restoration. Of all the composites, a microfill's shiny finish is most like enamel. Typically, they are not used in posterior restoration due to their lack of strength.

Microfill restoration frequently look so good that, even to a trained eye, it is hard to discern them from enamel or porcelain. Since their esthetic appeal can be compromised much more easily than porcelain, it is essential to take the utmost precautions in caring for them.

◆ ◆ ◆ ▬▬▬▬▬

MICROFILL AND
HYBRID
COMPOSITES
CHAPTER 6
PAGE 53

DIASTEMA CLOSURES

The closing of a diastema has been another advantage of microfills. The wide range of colors offered, as well as the surface finish, allow dentists to perfectly match the color of the surrounding enamel with a material that matches enamel's luster.

In the maintenance of these delicate procedures, it is important for the hygienist to recognize that an unusual emergence profile resulting in an excess bulk of composite material exists with most diastema closures. Unlike many other restorations, this excess bulk of material is usually unavoidable. The cervico-proximal line angle (mesial or distal, depending where the space was) is where you will usually find the beginning of this unusual emergence profile. Often the closure is detectable from the lingual aspect of the tooth. (Your magnification really comes in handy at moments like this.)

The margin of a diastema closure should always feel smooth to an explorer or curette. It is important not to roughen the margin or try to remove any overhangs. Alert your dentist if you have any questions or feel the margin could possibly be smoother. If there is persistent gingivitis around the tooth, the dentist should also be alerted so that the margin can be smoothed.

Closing of a diastema involves cantilevering composite material over tissue. Because of this cantilevering procedure, there is a tendency to create ledges at the cervical tooth-resin junction. These ledges must be eliminated when the diastema closure is originally done. If they are not, the probability of increased plaque build-up and subsequent gingival inflammation is increased.

The junction between the tooth and the restorative material should always be smooth to allow flossing to be done in a normal manner. (Shredding of the floss indicates additional smoothing of the restoration is necessary.)

━━━━━◆ ◆ ◆

MAINTAINING
ESTHETIC
RESTORATIONS
CHAPTER 6
PAGE 54

A Rota-dent (ProDentec) or Interplak (Bausch & Lomb) are excellent choices for patients to keep their diastema closure clean. Flossing should be taught with care, so that the patient can alert the dental office if the floss eventually begins to shred. Catching a rough margin may result in breaking or weakening the bond.

If the diastema is closed by only filling the space instead of veneering the entire tooth, curette the entire area and polish the enamel and the tooth with composite (aluminum oxide) polishing paste. Remember to carry the paste interproximally with the floss so you can polish and smooth the margin.

STAINING AROUND THE DIASTEMA

Aluminum oxide polishing paste typically does not remove the stain from a heavily stained composite. If the diastema has some stain that polishing did not remove, select a rubber polishing instrument or aluminum oxide disc. If the stain is at the gingival margin, retract the gum tissue with a plastic instrument. Interproximal stains need to be removed with finishing strips.

After the stain is removed, polish again with aluminum oxide polishing paste. (If some staining is still left, see aluminum oxide discs.)

CHIP REPAIRS

Microfills are also used to repair fractured anterior teeth. It is important to visualize the tooth-microfill margin and not to scratch or ditch this area. The margin is also a weak link of the restoration. Use very light pressure when polishing this area.

◆ ◆ ◆ ▬▬▬

MICROFILL AND
HYBRID
COMPOSITES
CHAPTER 6
PAGE 55

If part of the tooth is restored with a microfill composite and enamel is still present, treat the entire tooth as a microfill composite. Polish with aluminum oxide polishing paste. Use rubber polishing cups or aluminum oxide discs to remove excess stain.

DIRECT RESIN VENEERS

Direct resin veneers are hand-sculpted directly on teeth to be veneered, usually in one appointment. This procedure is not recommended for patients whose hygiene is not up to par, or for patients who cannot or will not commit to the recall program necessary for proper maintenance.

INDIRECT RESIN VENEERS

Indirect veneers are made in the laboratory. They require two appointments. Their advantage over direct veneers is the ability to make them less porous and more stain resistant. Patients may chose this procedure over porcelain because it is less expensive.

MACROFILL OR SMALL PARTICLE COMPOSITE

Macrofill composites, more commonly known as small particle, have two classes of filler particles. These restorations are not used very often today, but are still found in many patients' mouths. The large particles have an average filler size of 25 microns and the smaller particles have an average filler size of one to five microns. These composites are not very polishable and have generally been replaced by hybrids.

THE SUPERMAN OF PLASTICS: HYBRID COMPOSITES

Hybrids, or blend composite resins, are a marriage between the small particle and microfill composites. They are almost as polishable as microfills but have many physical properties of a macrofilled resin which imparts strength and durability

Hybrid composites are probably the most widely used for posterior restorations today. These materials have the advantage of being stronger than their cousin, microfill composites. In anterior restorations, hybrids are mainly used for dentin substitutes and may be covered with a microfill for attaining a higher shine. However, several manufacturers have improved their hybrid's surface finish to almost a microfil

◆ ◆ ◆

MAINTAINING
ESTHETIC
RESTORATIONS
CHAPTER 6
PAGE 56

uster. Due to this great advancement, these materials recently have been used in anterior restorations without being covered by a microfill.

Hybrid composites are, however, more frequently found in posterior restorations. This is one of the controversial issues in dentistry today, because most of the research done to date has shown composites in general do not hold up as well as amalgams in high-stress areas. Interestingly, research has also found composites strengthen weakened teeth when used with the right adhesive techniques.[7]

Despite the controversy, the fact remains patients love hybrid's esthetic appeal and many dentists are using them for Class I and II restorations that used to require amalgam.

ROUTINE CLEANING AND POLISHING OF MICROFILLS, HYBRIDS, AND SMALL PARTICLE COMPOSITES

Microfills, hybrids, and small particle composites are cared for exactly the same. It is important to understand that the surface finish of the three composites will not be the same. Microfills will always produce the highest luster followed by hybrid's glossy surface finish. Small particle composites finish to a much duller surface in comparison with microfills and hybrids.

◆ ◆ ◆

Clinical Tip: Always use 2 power (or greater) magnification lens when working with composites as well as for routine scalings and polishings. Magnification is essential in telling the difference in microleakage and stain along a margin of a restoration.

◆ ◆ ◆

◆ ◆ ◆ ▬▬▬

MICROFILL AND
HYBRID
COMPOSITES
CHAPTER 6
PAGE 57

To differentiate a porcelain veneer from a microfill veneer or enamel, gently run your explorer over the surface. The microfill veneer will feel softer than enamel's hard surface.

When an explorer glides over a porcelain veneer, a "scratchy feel" can almost be heard, which is unlike a microfill or enamel. A porcelain veneer's hardness mimics enamel and the surface of porcelain is much harder than a microfill surface.

Step 1: Curettes should always be used when scaling a tooth with any composite restoration. Scalers can scratch or ditch the materials. Curettes optimize esthetics leaving the restoration free of plaque and calculus.

Step 2: Use aluminum oxide polishing paste. Apply the paste directly on the restoration, making sure the entire restoration is covered.

Step 3: Put a small drop of water from the air/water syringe in the rubber cup before polishing. Polish each restoration for 30 seconds.

Step 4: With the paste still on the restorations, floss the teeth to carry the paste interproximally.

Step 5: Wash the paste, dry the teeth, and inspect the restoration. You are finished if there is no visible stain.

Step 6: Have the patient rinse for four minutes with sodium fluoride.

If there is stain on the margins or line angles after polishing with aluminum oxide paste, finishing strips and discs are then used to remove this stain.

FINISHING AND POLISHING STRIPS AND DISCS
Finishing and Polishing Strips

Proximal stain or roughness can be removed with finishing strips.

Step 1: *For light stain,* start with the fine grade strip. Carefully insert the non-abrasive middle plastic portion of the strip so he contact is not opened. (Epitex Strips, ICI/Coe do not have his non-abrasive center gap, but are very thin and excellent or this procedure.) Then use the super-fine strip to leave the restoration as smooth as possible. Carefully inspect the tooth. Go to step 2 if there is no visible stain.

or heavy stain, begin with the medium grade strip and graduate back up to the super-fine, leaving the restoration as smooth as possible. Carefully inspect the tooth. Go to step if there is no visible stain.

tep 2: Polish the proximal surfaces by applying the paste directly on the tooth, then wetting a rubber cup. Polish for 0 seconds. Carry the paste interproximally with floss.

inishing and Polishing Discs

Aluminum oxide polishing discs quickly and easily remove tenacious stain on composites. These discs come in four grades, but for stain removal use only the three finest: extrane, fine, and medium. Leave out the coarse grit for routine tain removal as it can easily change the contours of the restoration. A plastic instrument* should be used for gingival traction when necessary. These discs should be used only after the operator has been skillfully trained in their use.

A plastic instrument is a double ended metal instrument with two flat blunt ends at different angles to the handle. It was designed to place the original composites and was called plastic filling instrument. A fulcrum should always be used with the plastic instrument to prevent its slipping, which could possibly harm the tissue.

ep 1: *For medium to light stain*, begin with the fine polishing disc. Be sure not to use heavy pressure so you do not over-heat the restoration or the tooth. Carefully examine the tooth. Go to step 2 if there is no visible stain.

or heavy stain, begin with the medium aluminum oxide polishing disc. Polish for 10 to 15 seconds. Rinse the restoration thoroughly. If there is no visible stain, go to step 2.

♦ ♦ ♦ ▬▬▬▬

MICROFILL AND
HYBRID
COMPOSITES
CHAPTER 6
PAGE 59

Step 2: Use the fine aluminum oxide polishing disc. Carefully polish the same surface polished by the medium disc for 10 to 15 seconds.

Step 3: The extra-fine disc is now used in the same manner as the fine disc. Go over the tooth with an explorer to ensure the tooth is left as smooth as possible. Rinse thoroughly.

Step 4: Thoroughly polish the previously stained area with aluminum oxide polishing paste for 30 seconds. Rinse thoroughly.

RUBBER POLISHING INSTRUMENTS

Composite rubber polishing instruments were designed to smooth and polish microfills and hybrids, as well as remove light stain. They incorporate aluminum oxide in a soft rubber and are available in a variety of shapes, adapting easily to every surface. They save an extra step (compared to the discs) since they usually come in only two grits, finishers and polishers. The finishers smooth the restoration and remove light stain. The polishers leave the restoration as smooth as possible.

Step 1: For occlusal surfaces and embrasures, use a pointed or pear shaped finisher. Next, use the polishing point, leaving the restoration as smooth as possible. Use a light, uniform stroke. Too much pressure can overheat the restoration and the tooth. Rinse thoroughly. Carefully examine the tooth to ensure a smooth surface free of stain.

◆ ◆ ◆

MAINTAINING
ESTHETIC
RESTORATIONS
CHAPTER 6
PAGE 60

Step 2: For facial surfaces, use the cup-shaped finisher, then polisher instruments. Cups can also be used on the lingual surfaces of posterior teeth.

Step 3: Polish again for 30 seconds with the aluminum oxide polishing paste and rinse well with water to make sure any residue from the rubber polishing instruments is removed.

DAMAGE TO COMPOSITES

Studies have shown the initial smooth surface of all composites can be jeopardized by the subsequent use of common cleaning devices found in most dental offices.

1. Coarse prophy paste may roughen and scratch the surface of composites.[1]

2. Sonic or ultrasonic scalers can scratch the surface, "pluck" out filler particles, and possibly even weaken the bond of a composite.[2]

3. Air-powder abrasive instruments (Prophy Jet, Dentsply) can scratch and pit the surface of composites, and even remove part of the composite material.[3] (See photographs)

EFFECT OF A HYDRAULIC JET PROPHYLAXIS SYSTEM ON COMPOSITES

This study tested the effects of a hydraulic jet prophylaxis system on a total of twenty-five simulated composite restorations in Ivorine blocks. Microfill, small particle, and large particle restorations were used. The surface of each restoration was polished flat and flush with its respective Ivorine block. A Prophy-Jet (Dentsply) was placed 4 mm above each specimen's surface and applied with a 6 mm range in a back and forth motion for 20 seconds. A hybrid composite was not used in the study, however, notice that both the Prisma-fil (small particle) and Silux (microfill are noticeably changed after the 20 seconds exposure. Since hybrids are a mix of small particle and microfills, this is perhaps a good indication of the expected result. Even Heliomolar, a heavily filled microfill, showed a damaged surface caused by 20 second exposure (15 cycles) by the Prophy-Jet.

◆ ◆ ◆ ▬▬▬

MICROFILL AND
HYBRID
COMPOSITES
CHAPTER 6
PAGE 61

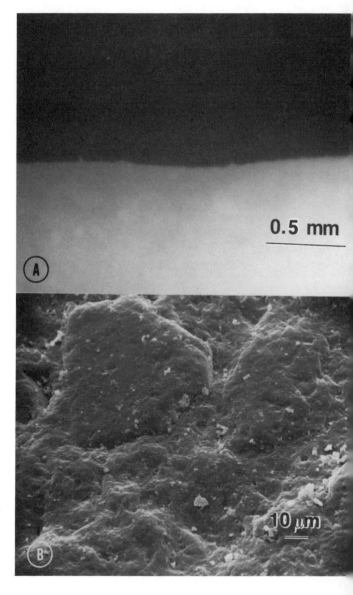

Figure 1: A, PHOTOMICROGRAPH OF CROSS SECTION
OF ABRADED SILUX COMPOSITE. B, SEM MICROGRAPH
OF ABRADED SURFACE IN SILUX COMPOSITE.

*Courtesy of Dr. C.R. Reel and the Journal of Prosthetic
Dentistry.*

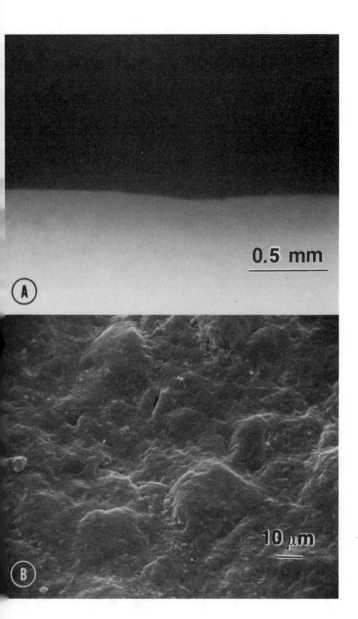

0.5 mm

Ⓐ

10 μm

Ⓑ

◆ ◆ ◆ ▬▬▬▬

MICROFILL AND
HYBRID
COMPOSITES
CHAPTER 6
PAGE 63

Figure 2: A, PHOTOMICROGRAPH OF CROSS SECTION
OF ABRADED HELIOMOLAR COMPOSITE. B, SEM
MICROGRAPH OF ABRADED SURFACE IN HELIOMOLAR
COMPOSITE.

Courtesy of Dr. C.R. Reel and the Journal of Prosthetic
Dentistry.

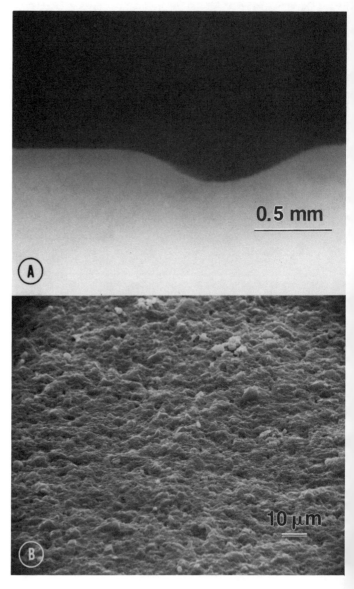

Figure 3: A, PHOTOMICROGRAPH OF CROSS SECTION
OF ABRADED PRISMA-FIL COMPOSITE. B, SEM
MICROGRAPH OF ABRADED SURFACE IN PRISMA-FIL
COMPOSITE.

*Courtesy of Dr. C.R. Reel and the Journal of Prosthetic
Dentistry.*

4. A hard toothbrush and abrasive toothpaste may help to accelerate further breakdown of the composite by also increasing surface roughness and abrasion of the material.[4]

The frequent use of any of the above mentioned devices may ultimately cause increased surface roughness that may collect debris and plaque. This can lead to gingival inflammation, staining, recurrent caries, and/or accelerated wear.[5]

Choose to be a Top Performer!
We are Free up to the Point of Choice, Then
the Choice Controls the Chooser.

References:

(1) de Wet, FA; Ferreira, MR, Polishing procedures for microfilled resins. Journal of the D.A.S.A. 1982; 37,797-803.

(2) Zitterhart, P.A., Effectiveness of ultrasonic scalers: A literature review, General Dentistry/July-August 1987, pp 295-297.

(3) Reel, C.R.; Abrams, H.; Gardner, S.; Mitchell, R. J., Effect of a hydraulic jet prophylaxis system on composites. The Journal of Prosthetic Dentistry 1989; 61:441-5.

(4) Strassler, H.E.; Moffitt,W., The Surface Texture of Composite Resin after Polishing with Commercially Available Toothpastes, Compendium of Continuing Education in Dentistry, Vol. VIII, No. 10, pp.826-830.

(5) van Dijken, J.M.V.; Ruyter, I.E., Surface characteristics of posterior composites after polishing and toothbrushing. Acta Odontol Scand 1987; 45:337-346.

(6) Reality, Vol. 4, No.1, Anterior Hybrids/ Small Particle Composites, Page 10.

(7) Reality, Vol. 4, No.1, Posterior Composites, Page 71.

◆ ◆ ◆ ▬▬▬

MICROFILL AND
HYBRID
COMPOSITES
CHAPTER 6
PAGE 65

■■■■■■ ◆ ◆ ◆

MAINTAINING
ESTHETIC
RESTORATIONS
CHAPTER 6
PAGE 66

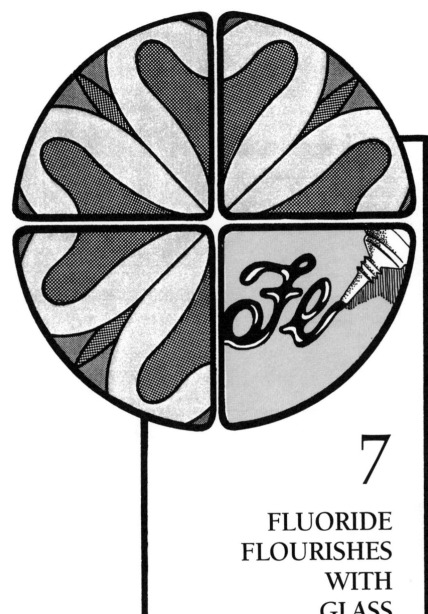

7

**FLUORIDE
FLOURISHES
WITH
GLASS
IONOMERS**

■■■■■■■ ◆ ◆ ◆

MAINTAINING
ESTHETIC
RESTORATIONS
CHAPTER 7
PAGE 68

CHAPTER 7

◆ FLUORIDE FLOURISHES WITH GLASS IONOMERS

Glass ionomers are used for Class III and V restorations, as well as bases and liners underneath restorations and for cementing crowns and bridges. As they release fluoride, glass ionomers are especially indicated for patients with a high caries level. But they have limited appeal because their esthetics is not equal to composite.[1]

Glass ionomers have another distinct disadvantage. If they dry out they can start cracking, causing microleakage and failure of the restoration.

Glass ionomers should be cleaned and polished like a composite. (See Chapter 6 on how to care for composites.) Special care should be taken not to desiccate these restorations during polishing procedures.

Since glass ionomers are often placed in elderly patients, who have an increased risk of periodontal disease, it is important that patients take extra care during their oral hygiene regimen to keep from abrading these restorations. A soft or electric toothbrush is especially useful in these cases. A Rota-dent (ProDentec) or Interplak (Bausch & Lomb) are excellent choices for cleaning these restorations. Hydrogen peroxide and/or a neutral sodium fluoride are the best home rinses.

Due to decreased dexterity and gingival recession associated with periodontal disease, elderly patients sometimes have a more difficult time with home hygiene. To decrease the susceptibility to root caries and periodontal disease, a two month recall is advised.

FLUORIDE
FLOURISHES
WITH GLASS
IONOMERS
CHAPTER 7
PAGE 69

◆ ◆ ◆

WHEN POLISHING GLASS IONOMERS, IT IS NECESSARY TO COAT THE RESTORATION WITH COCOA BUTTER OR VASELINE TO PREVENT DESICCATION.

◆ ◆ ◆

ROUGHNESS OR STAIN ON THE FACIAL OR LINGUAL SURFACES:

Step 1: Select a rubber finishing cup or point. Cups are best for broad facial surfaces. Points are best for root surfaces, lingual cingulums, line angles and proximal surfaces where cups cannot reach.

Gently "polish" out the roughness, using light strokes to prevent overheating the restoration or the tooth. Gently glide an explorer over the restoration, making sure it is smooth. Rinse thoroughly. If the restoration is smooth and free of stain, go to step 2.

Step 2: Graduate to the polishing cup or point. Apply light uniform strokes being careful not to ditch the restoration.

Step 3: Apply a generous amount of aluminum oxide polishing paste so that it completely covers the restoration. Wet the prophy cup with water, and polish for 30 seconds. With the paste still on the tooth, carry the floss over the line angles and proximal surfaces where the rubber points and cups have been used.

Step 4: Apply sodium fluoride for one minute.

ROUGHNESS ON THE PROXIMAL SURFACES

Step 1: With the finest grade of finishing strip, insert the middle non-abrasive plastic portion of the strip so the restoration or the tooth is not abraded.

Step 2: Use a back and forth motion to smooth out the rough area of the restoration. With an explorer, check the restora-

■■■■◆ ◆ ◆

MAINTAINING
ESTHETIC
RESTORATIONS
CHAPTER 7
PAGE 70

tion. If it is still not smooth, graduate to the next coarsest grade. Continue this process, until the restoration feels smooth to the explorer.

Step 3: Graduate back to the finest strip to ensure the smoothest possible surface.

Step 4: Polish with aluminum oxide polishing paste. With the polish still on the tooth, carry it carefully over the area that has just been smoothed with the aluminum oxide strips.

Step 5: Carefully inspect the restoration. You are finished if the tooth is smooth and free of stain.

Step 6: Apply sodium fluoride for four minutes.

Glass Ionomers are soft materials and are affected by abrasives used in dental treatment. Two such abrasives are the air-powder abrasive spray and prophy paste. Both techniques are equally effective in removing stain. However, significant surface roughness has been shown with composites and glass ionomers after their use.[2]

Ultimately, it is important to remember that glass ionomers are easily desiccated and it is wise to not only establish safe clinical techniques, but also to ensure patients are not taking medication which could cause xerostomia. Except for the desiccation factor, all of the maintenance rules that apply to composites should also be used with glass ionomers. (See chapter 6.)

FLUORIDE
FLOURISHES
WITH GLASS
IONOMERS
CHAPTER 7
PAGE 71

Today,
If the Season in One of Your Patient's Life is
Fall, Make it Rich with Color. Smile with all
Your Heart.

References:

1. Hicks M. J.; Flaitz, C M.; Silverstone, LM, Secondary caries formation in vitro around glass ionomer restorations, Quintessence International, Vol. 17, Number 9,1986, pp.527-532.

2. Cooley RL; Brown, FH; Stoffers KW, Effect of air-powder abrasive spray on glass ionomers, AM J Dent 1:209-213, 1988.

■■■■■■ ◆ ◆ ◆

MAINTAINING
ESTHETIC
RESTORATIONS
CHAPTER 7
PAGE 72

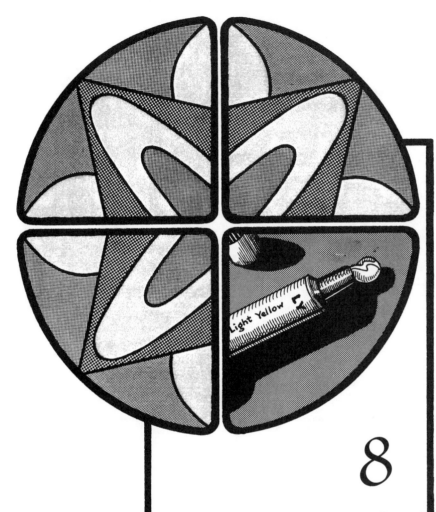

8

HOW TO
DEAL WITH
DEPOSITS
ON
COMPOSITES

CHAPTER 8

◆ HOW TO DEAL WITH DEPOSITS ON
COMPOSITES;
(Maintenance three to five years later)

Many patients have received direct and indirect resin restorations in the last five to six years. Unfortunately, because of the material used and the care by the patient and the dental team, these restorations may not look as beautiful today as they once did.

Even though these restorations may be aged, we have found many patients may resist having them replaced only for esthetics. (Note: This assumes that there is no recurrent caries or other functional problems with the restorations.) This chapter is devoted to updating aged resin restorations.

Over a period of years, resin materials suffer from discoloration and pitting. This is a result of filler particles being dislodged from the resin matrix and the resin matrix itself absorbing substances which can cause staining, such as coffee, tea, red wine, etc. And, as previously mentioned, resin materials can have small air pockets, even though manufacturers are vacuum packing them in their delivery syringes.

As a restoration ages, its surface wears. Small voids that originally were subsurface are now exposed to the oral environment. These small voids, if left unattended, can become larger voids due to the degradating effects of food, liquids for drinking and rinsing, plaque, etc. It becomes necessary to break the cycle and refine the restoration as closely back to its former smoothness and luster as possible.

◆◆◆ ▬▬▬

HOW TO DEAL
WITH DEPOSITS
ON COMPOSITES
CHAPTER 8
PAGE 75

There should be an understanding and comfort level established between the dentist and hygienist before this procedure is performed. Exactly how much enhancement to the
restoration the hygienist will be able to do should be discussed before this procedure begins. No one knows more
about the restorations than the person who created them.

If the patient's restorations were placed by another dentist,
the dentist who did the restorations needs to be consulted,
if possible. This will allow you to ascertain information such
as, for example, the thickness of veneers, shade, material,
whether the restorations were layered (hybrid + microfill),
etc. The original study models, x-rays, and photographs are
also helpful.

Before the Patient Arrives:

Step 1: Review the study models and original color slides or
photographs. If the dentist knows how much enamel was
removed before placing the restoration or veneer, it will help
you to determine how thick the veneers are.

Step 2: Discuss with your dentist the materials used during
original placement. For instance, if a a microfill was layered
over a hybrid you would not want to polish through the
microfill exposing the hybrid. This would preclude the possibility of re-establishing a high shine on the restoration.
Various shades and types of microfills are often used on the
same restoration.

Microfills are offered in both opaque and translucent finishes
and sometimes are layered on the tooth, depending on the
result the patient and the dentist want. Often the cervical
area of a restoration or veneer is a different shade from the
incisal, using two different microfills. Alteration of the original
"look" of the restoration is possible without the correct
knowledge. The procedure described below is for updating
and enhancing composite restorations, not altering the
original result.

"Updating" Worn Resin Veneers and Composite Restorations

The rough outer surface of the composite must first be smoothed and all overhangs removed. It is possible to refine and repolish resin to its original finish, whereas once the glaze of porcelain has been altered you will never completely restore it.

Materials necessary to begin enhancement procedure:

Latch-key Slow Speed Handpiece
Polishing disc~ medium, fine, extra-fine (small discs)
Polishing str.
Plastic Instrument
Composite rubber polishing instruments
Prophy cups
Composite polishing paste

Appointment Time: one hour for six veneers or composite restorations

Charge: $60 - $100

Step 1: REMOVE EXTRINSIC STAIN: Clean and polish teeth with aluminum oxide polishing paste and a webbed rubber prophy cup to remove external stains, etc. Gingival tissue should be healthy with no bleeding present.

Step 2: REMOVE SURFACE ROUGHNESS, CRACKS, AND INTRINSIC STAIN: Insert a medium grit finishing disc (small) into the latch-key handpiece. This disc will remove the outer chipped and stained layer of composite. Begin at the cervical margin. Place the blunt end of the plastic instrument gently on the tooth just apical to the cervical margin. This will retract and protect the tissue from the disc. The disc should be straight and parallel to the angle of the tooth being refined, so that it does not ditch the composite. If you have the original material in the office, compare the color of the present restoration to the original color. (Check in the chart for the original color).

♦ ♦ ♦ ▬▬▬

HOW TO DEAL
WITH DEPOSITS
ON COMPOSITES
CHAPTER 8
PAGE 77

Begin by finishing each section of the restoration with the medium disc for 10 seconds. For heavy stain or abrasion, it might be necessary to change the disc several times. Clean and dry the tooth, inspecting it for remaining abrasion or stain. Compare the present color to the original color. You are ready to go to the next step if the original and present colors match and if the tooth is smooth and stain free. NOTE: Some composites will change color over time and no amount of finishing and polishing will get the restoration to match new material.

Step 3: PROXIMAL STAIN: Start with a medium finishing strip. Using the middle non-abrasive plastic portion, insert it carefully between the teeth to prevent opening a contact. Work the strip in a back and forth motion on the stained portion of the restoration for 10 seconds. Repeat this step until the restoration is smooth and free of stain. Graduate to the finest strip to ensure the restoration is left as smooth as possible.

Step 4: POLISHING I: Using a fine aluminum oxide polishing disc parallel to the tooth, begin polishing the cervical portion of the restoration. Polish each section of the restoration with the disc for 10 seconds. Softly run an explorer over the restoration making sure it is smooth.

Step 5: POLISHING II: Using an extra-fine aluminum oxide polishing disc parallel to the tooth, begin polishing the cervical portion of the restoration. Polish each section of the restoration with the disc for 10 seconds. Remove polishing dust and debris carefully with air. Check to be sure all voids and stain have been removed.

Step 6: POLISHING III: Insert a webbed rubber prophy cup into the slow handpiece. Cover each restoration with a generous amount of aluminum oxide polishing paste. Place a small drop of water into the prophy cup. Polish each tooth for 30 to 45 seconds. With floss carry the paste still on the restoration interproximally. Floss in a shoe-shine like manner. Rinse, dry and inspect the restoration.

Now that you have recreated a beautiful restoration, please give the patient a "Caring for you Cosmetic Restoration Sheet", an ultra-soft toothbrush, and a tube of non-abrasive toothpaste.

Expect the Best From Yourself and Give Your Best to Others!

◆ ◆ ◆ ▬▬▬▬

HOW TO DEAL
WITH DEPOSITS
ON COMPOSITES
CHAPTER 8
PAGE 79

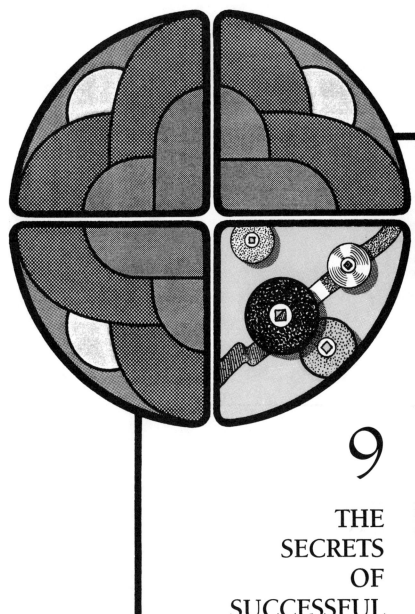

9
THE
SECRETS
OF
SUCCESSFUL
POLISHING

■■■◆ ◆ ◆

MAINTAINING
ESTHETIC
RESTORATIONS
CHAPTER 9
PAGE 82

CHAPTER 9

♦ THE SECRETS OF SUCCESSFUL POLISHING

Correctly polishing dental restorations is vital in maximizing the esthetic, as well as the functional life of the restoration. Polishing all the dentition for its esthetic appeal is well-known to both patients and dental teams. Polishing's functional assets to remove plaque and minimize gingivitis, as well as increase the life of restorations, is universally known to both hygienists and dentists.

The functional and esthetic aspects of any restoration are interdependent of one another. A well preserved, esthetically pleasing restoration is not only free of stain with a smooth lustrous surface, it also does not collect debris and bacteria. Functionally, if a restoration is kept smooth and clean, the tooth will more likely resist plaque. A rough surface on any restoration may promote gingival inflammation, staining of the filling, recurrent caries, and abrasion.[1]

The polishing of microfill and hybrid composites, resin cements, and porcelain has taken on an entirely new complexion. As we have proven in the past chapters, the polishing of cosmetic restorations can be a capricious animal. Just when you think you are helping the restoration, you find you have actually helped to deteriorate it!

Composite and porcelain restorations becomes less lustrous over time due to acids in foods, abrasive brushes, coarse prophy paste, acidulated phosphate fluoride, and plaque. The loss of luster is due to large scratches in the surface material. To provide our patients with the most beautiful restorations

♦ ♦ ♦ ▬▬▬▬

THE SECRETS OF
SUCCESSFUL
POLISHING
CHAPTER 9
PAGE 83

possible, we must turn these larger scratches into smaller and smaller scratches.[2]

Aluminum oxide discs and strips, as well as rubber polishing instruments, have proven to be the greatest resources for esthetic restorations. Depending on the degree of wear on each restoration, these instruments will solve almost any composite or porcelain restorations' polishing need.

As in all areas of our lives, experience is the best teacher. Over time, it will become very easy to quickly judge which instrument is best for "that" restoration's particular problem. The one rule that should always be followed:

DO NOT RESORT BACK TO COARSE PROPHY PASTE, HAND SCALERS, SONIC OR ULTRA-SONIC SCALERS OR AIR-POWDER ABRASIVE INSTRUMENTS.

POSTERIOR RESTORATIONS

Bacterial plaque accumulates faster in the posterior region of the mouth. In addition, Class II restorations are more difficult to clean than anterior restorations, both for the patient and hygienist. Therefore, to protect the restoration and provide it with the longest life possible, it is necessary to routinely polish porcelain and composite posterior restorations with rubber polishing instruments to remove any rough areas before using polishing paste. This is necessary even if there is no stain present.

Porcelain in posterior restorations

Porcelain is much harder and more abrasive than enamel. When placed in a posterior restoration, porcelain has the potential to abrade the opposing teeth. Therefore, it is very important to keep porcelain inlays, onlays, and crowns well polished. The cusps of porcelain restorations should be free of rough edges, which would increase further wear to the opposing dentition. Should a cusp become jagged or rough, a finishing diamond must be used to smooth it followed by polishing with rubber instruments and diamond paste. Consult your dentist if a rough or questionable surface becomes noticeable during a routine cleaning appointment.

Once a year, it may be necessary to "polish" posterior porcelain restorations with porcelain rubber polishing instruments to ensure their smoothness. See "Finishing with Rubber Polishing Instruments for Composites and Porcelain Posterior Restorations".

Composites in posterior restorations

Posterior composite restorations are kinder than porcelain to the opposing teeth since they are softer than enamel. As we have mentioned earlier, composites can wear and become scratched much more easily than porcelain.

Because of their "softness", posterior composites can wear even more quickly than anterior composites. Therefore, only finish these restorations when absolutely necessary to remove adherent stain and roughness to preserve their original contours as much as possible.

HEAVY STAIN REMOVAL FOR COMPOSITES AND PORCELAIN RESTORATIONS WITH ALUMINUM OXIDE DISCS

Aluminum oxide discs come in two sizes, large and small, and four grits, coarse, medium, fine, and extra-fine. The large size and the coarse grit should not be used for this procedure. The operator should be carefully trained in the use of discs before this procedure is implemented.

Aluminum oxide discs should be used with light even pressure. Too much pressure can abrade the restoration and overheat the tooth.

Step 1: For heavy stain, begin with a small, medium grit aluminum oxide disc. For stain at the cervical margin, use a plastic instrument to protect the tissue from the disc. Polish the stained area for 10 to 15 seconds.

Step 2: For light stain, begin with the small, fine grit aluminum oxide disc. Protect the tissue with the plastic instrument. Polish the stained area for 10 to 15 seconds. (A plastic instrument is a double-ended metal instrument with two flat blunt ends. It was originally designed to place composites and thus was called a plastic filling instrument.)

Step 3: Remove polishing dust with air and inspect the tooth. If the stain is gone, use the extra-fine polishing disc for 10 to 15 seconds, or the same amount of time used with the previous disc. If there is still some stain, repeat steps 1 or 2. Gently glide an explorer over the polished area to ensure the restoration is smooth.

Step 4: Polish. (see Final Polishing below)

FINISHING WITH RUBBER POLISHING INSTRUMENTS FOR COMPOSITES AND PORCELAIN POSTERIOR RESTORATIONS

Rubber polishing instruments are available for both porcelain and composites. The porcelain rubber polishing instruments incorporate silicone dioxide in the rubber. For composites, aluminum oxide is incorporated into the rubber.

Rubber polishing instruments should always be used dry in the slow speed handpiece. Light, even pressure is necessary to prevent overheating the restoration or the tooth.

■■■■■◆ ◆ ◆

MAINTAINING
ESTHETIC
RESTORATIONS
CHAPTER 9
PAGE 86

Step 1: OCCLUSAL SURFACES: A pointed or pear shaped rubber polishing instrument is best to get in the fissures and curves of the occlusal surfaces. If the surface is stain free and relatively smooth, use the "polisher." If there is stain or roughness, use the "finisher", graduating up to the polisher. Rinse well for 30 seconds. Dry the restoration, and examine carefully. You are through if there is no visible roughness or stain.

Step 2: BUCCAL, FACIAL, AND LINGUAL SURFACES: Rubber polishing instruments in the shape of cups or discs are best for buccal and facial surfaces. Points are best for ngual surfaces. If the margin is smooth and the restoration stain free, begin with the "polisher". If there is stain or roughness at the margin, select the "finisher", graduating back up to the "polisher."

tep 3: PROXIMAL SURFACES: Finishing strips are necessary to remove stain and to smooth these surfaces. If there is tain, begin with the medium strip. Carefully insert the middle on-abrasive plastic portion so a contact is not opened. Epitex polishing strips (ICI/Coe) do not have a non-abrasive middle portion.) Polish in a back and forth motion for 10 econds. Graduate to the fine and extra-fine strips, leaving the restoration as smooth as possible. If the restoration is tain-free and feels smooth to the explorer, begin with the ne strip, graduating up to the extra-fine.

OTE: Some contacts between teeth are so tight they preude passing a strip betweem them. In this case, use a narrow rip inserted apical to the contact. This is similar to using a oss threader with a fixed bridge. With this technique, you can mooth and polish the gingival area of the proximal surce which is usually where you find most stain and roughness.

he Profin Directional System (Weismann Technology) is an xcellent attachment for your straight handpiece. When acvated, this instrument moves its specially designed tips back nd forth. Because of this back and forth motion, you can ain access into areas such as gingival embrasures to remove verhangs or tenacious stain on composites or porcelain.

tep 4: POLISH. (see Final Polishing below)

LWAYS RINSE THE TOOTH THOROUGHLY (30 ECONDS) AFTER USING ALUMINUM OXIDE STRIPS, LUMINUM OXIDE DISCS, AND RUBBER POLISHING OINTS

his ensures all small particles from the restoration or the olishing instrument will not attract plaque and create future oughness.

FINAL POLISHING

Step 1: Dry the restorations. Apply a generous amount of paste (aluminum oxide for composites and diamond for porcelain) to the restorations. For composites, place a small drop of water in a prophy cup and shake out the excess water. Polish each restoration for 30 to 45 seconds.[3]

◆ ◆ ◆

Clinical Tip: APPLYING COMPOSITE POLISHING PASTE DIRECTLY TO THE TOOTH IS THE SECRET TO SUCCESS FOR A LUSTROUS FINISH ON ALL COMPOSITES.

◆ ◆ ◆

Step 2: With the paste still on the tooth, floss each tooth, carrying the paste interproximally. Floss the proximal surfaces with shoe-shine like strokes.

Step 3: Rinse thoroughly. Aluminum oxide and diamond polishing pastes are tenacious and sometimes difficult to get off the teeth.

Step 4: Apply sodium fluoride for four minutes.

You Can Have Everything You Want in Life, if You Will Help Enough Other People Get What They Want.

■ ◆ ◆ ◆

MAINTAINING
ESTHETIC
RESTORATIONS
CHAPTER 9
PAGE 88

References:

1. van Dijken J.W.V.; Ruyter,E.I., Surface characteristics of posterior composites after polishing and toothbrushing, ACTA Odontal Scand, 1987;45:337-346.

2. Christensen, R.P.; Christensen, G.J., Comparison of Instruments and commercial pastes used for finishing and polishing composite resin, General Dentistry, Jan.-Feb.1981, pp.40-45.

3. REALITY, Vol. 4, #1, Polishing Composites, page 126.

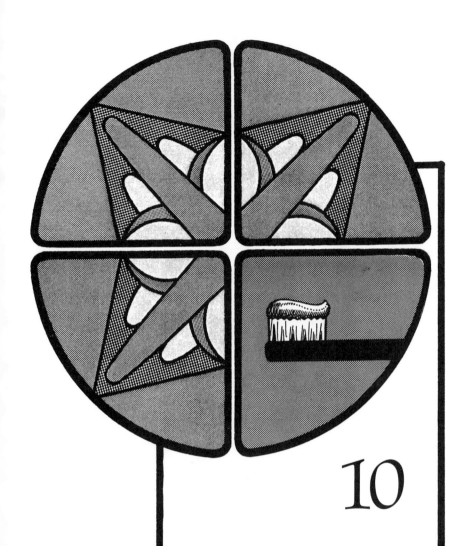

10

**DON'T LET
HOME CARE
BECOME YOUR
NIGHTMARE**

CHAPTER 10

◆ **DON'T LET HOME CARE BECOME YOUR NIGHTMARE!**
TOOTHPASTES, FLUORIDES, AND HOME CARE

Have you ever seen a dentist work very hard to create a beautiful restoration and mysteriously, one or two years down the road, it begins to fail? The failure could be attributed to the patient's home care.

A CASE REPORT

Not too long ago, a beautiful young lady came to our office. She had been in a terrible car accident two years before and her maxillary centrals, being broken in the accident, had been repaired with microfill composite restorations. This young women asked if there was anything that could be done to remove the dark brown stain and improve the dull restorations that, not too long ago, had been beautiful. We asked her about her routine home care. She proudly proclaimed that she brushed three to four times a day with baking soda and a hard toothbrush and then rinsed with Listerine. (She had read about the benefits of baking soda for the "gums" in a health magazine.)

It would be impossible for anyone to know what her composite restorations would look like today if she had cared for them properly. However, it would be safe to speculate that the microfill material would not have deteriorated quite so rapidly with proper care.

◆ ◆ ◆ ▬▬▬

DON'T LET HOME
CARE BECOME
YOUR NIGHTMARE
CHAPTER 10
PAGE 91

TOOTHBRUSHES AND TOOTHPASTES FOR PORCELAIN AND COMPOSITES

Toothbrushes

Soft toothbrushes are now the professional standard due to their ability to thoroughly clean without abrasion to the teeth or gingiva. Even though medium or hard toothbrushes are not recommended by most dental professionals, many are still sold each year, and some patients love them. To our knowledge, a study has not been done on the effects of hard toothbrushes on porcelain and composites. However, in lieu of the fact composites and resin cements are softer than enamel and porcelain has a glaze that can be abused, hard or medium toothbrushes are not recommended.

An ultra-soft toothbrush is recommended for all cosmetic restorations. The soft bristles can gently reach into the cervical area between a composite or porcelain restoration and the gingival tissue, thoroughly eliminating plaque and debris without harming the restoration. Rota-dent (ProDentec) or Interplak (Bausch & Lomb) electric toothbrushes have very soft bristles that efficiently clean composite and porcelain restorations, as well as massage the gingiva.

Toothpastes

A recent study evaluated the surface of composite resins following the use of various commercially available toothpastes. The study found that rough composite surfaces can contribute to stains, recurrent caries, and plaque and food retention. In addition, the patients in the study often complained that their restorations felt rough, as if the composite was unpolished, after using abrasive toothpastes.

One of the most important findings of the study was that toothbrushing alone was not sufficient for removing pellicle and stain. The dentifrice must contain the necessary abrasive to remove stains, assist in the removal of plaque, and introduce fluoride to the tooth surfaces and to other intraoral tissues.[1]

◆ ◆ ◆

MAINTAINING
ESTHETIC
RESTORATIONS
CHAPTER 10
PAGE 92

Toothpastes found to be most abrasive to composite resins are Advanced Formula Crest (Proctor and Gamble), Colgate (Colgate-Palmolive), Aqua-fresh (Beecham), Pearl Drops (Carter Products),and Topol(Jeffrey Martin). The next roughest toothpaste was Crest Tartar Control Formula (Proctor and Gamble), while the smoothest was Rembrandt (Den-Mat) toothpaste.[1]

Previous studies reviewed in this book reveal the abrasivity of baking soda.Whether the baking soda is in an air-powder abrasive instrument or on a toothbrush, it is not recommended for composite or porcelain veneers. The configuration of sodium bicarbonate crystals can significantly pit and dull cosmetic restorations and the glaze of porcelain.[2,3]

One of the keys to toothpaste acceptability for composite or porcelain is the metallic abrasive. Rembrandt (Den-Mat) toothpaste uses alumina, which is a microfine abrasive recommended for polishing composite resins, while Advanced Formula Crest incorporates titanium dioxide, an extremely hard metal.[1] EpiSmile (Epi Products) takes another approach to safeguarding cosmetic restorations. It cleans bonding or porcelain by bleaching out any residues from stains caused by food or tobacco. The agent responsible for this bleaching action is calcium peroxide, which also whitens enamel.[6]

NOTE: A new study[5] showed the differences in abrasiveness of toothpastes may not be as significant as previously thought. But, to be on the safe side, it is probably better to use either Rembrandt (Den-Mat) or EpiSmile (Epi Products) until further research is done.

FLUORIDES

Sodium fluoride (NaF) is the only known safe fluoride for cosmetic restorations. It is important that patients are made aware of the harm acidulated phosphate fluoride (APF) or stannous fluoride can do to their cosmetic restorations. Like our "case study", many patients are cleaning their teeth with something they read about in a magazine, or perhaps from a friend who happens to have something "on hand". Stannous

♦ ♦ ♦ ▬▬▬

DON'T LET HOME
CARE BECOME
YOUR NIGHTMARE
CHAPTER 10
PAGE 93

Even though APF is not found in many home rinses or gels, it is important the patient is made aware of the effects on cosmetic restorations. If, by chance, they get their teeth cleaned at another dental office by a hygienist unaware of APF's effects on cosmetic restorations, this knowledge would be very helpful. (It is also a great help to your practice.)

Acidulated phosphate fluoride's popularity is due to its ability to inhibit carious lesion formation and remineralize teeth. APF contains about 37% more fluoride than sodium and has a low pH, both of which favor greater fluoride deposition in carious lesions. A recent study comparing the two fluorides showed that sodium fluoride provided 74% protection from mineral loss in lesion formation. APF provided 100% protection. However, it was further observed that more frequent applications of sodium fluoride would provide clinical effects equal to APF.[4]

APF is still the agent of choice for many situations but the evidence clearly shows that frequent use of neutral sodium fluoride also is able to provide an important anti-caries effect.

Therefore, patients with a high caries index and cosmetic restorations, should have sodium fluoride applied for one minute at the end of their recall visit. Prescribe a neutral sodium fluoride rinse such as Act Fluoride (J & J) to be used daily at home.

PATIENT'S HOME CARE RESPONSIBILITY

Remember, when people make one lifestyle change, they will frequently make others. Cosmetic dentistry's home care program also falls into this category. The good news for the patient and the dental team is that the timing is excellent for establishing superior oral hygiene. It is also mandatory to keep their restorations looking terrific.

Education is the most motivating tool we can use to initiate and keep good habits. The greater understanding the patient has of the material in their cosmetic restorations and what factors will help or harm it, the better they will follow a conscientious home hygiene schedule.

RESPONSIBILITY TO THE PATIENTS

The cervical area of a veneer is the most susceptible to plaque, bacteria, and gingival inflammation. It is prudent to show the patient this area and make your recommendation for cleaning it while the patient is looking in the mirror.

Facts the patient should know:

Composite restorations are more susceptible to plaque and stain than porcelain or enamel.

Porcelain veneers can and will stain without adequate home care or frequent recalls. The resin cement can stain very easily. (Show the patient the location of the resin cement.)

Periodontal problems can still occur with composite or porcelain veneers, particularly when home care or a frequent recall protocol is not maintained.

If gingival bleeding occurs around a cosmetic restoration, the patient should make an appointment for a cleaning and to have the restoration carefully checked.

Flossing is not an option, but a necessary "habit" with cosmetic restorations.

Three to four month recall appointments are necessary with cosmetic restorations.

Ultra-soft toothbrushes are necessary with cosmetic restorations.

Rembrandt or EpiSmile are the only toothpastes recommended at this time.

Mouthwashes high in alcohol soften bonding and the resin cement around veneers.

◆ ◆ ◆ ▬▬▬▬

DON'T LET HOME
CARE BECOME
YOUR NIGHTMARE
CHAPTER 10
PAGE 95

Do not let another hygienist or dentist use prophy paste, ultra-sonic or sonic scalers, Prophy-Jets or APF fluoride.

Cosmetic Dentistry is changing rapidly. Dentists' and hygienists' ultimate responsibility is to keep informed of its changes. What is correct today, might be history tomorrow.

If Your Patient Does Not Look Good, You Do Not Look Good

. . . SO EDUCATE NOW, SMILE LATER.

References:

1. Strassler, H.E.; Moffitt, W., The Surface Texture of Composite Resin After Polishing with Commercially Available Toothpastes. Compend Contin Educ. Dent. Vol. VIII, No. 10, 1987, pp.826-830.

2. Barnes, Caren M.; Hayes, E.F.; Leinfelder, K.F., Effects of an airabrasive polishing system on restored surfaces, Gen. Dent. May/June 1987, pp.186-189.

3. Cooley, R.L.; Lubow, R.M.; Brown, F.H., Effect of air-powder abrasive instrument on porcelain, The J. Pros. Dent., Vol.60, No.4, Oct.1988. pp.440-443.

4. Melberg, J.R.; Lass A.; Petrou I.; Inhibition of artificial caries lesion formation by APF and neutral NaF office gels, Am J Dent 1:255-257,1988.

5. Goldstein, G.R.; Lerner, T., The Effect of Toothpastes on a Hybrid Composite Resin, AADR Abstract #442, San Francisco, 1989.

■◆ ◆ ◆
MAINTAINING
ESTHETIC
RESTORATIONS
CHAPTER 10
PAGE 96

11

A JOURNEY
TO
EXCELLENCE

■■■◆ ◆ ◆

MAINTAINING
ESTHETIC
RESTORATIONS
CHAPTER 11
PAGE 98

CHAPTER 11

◆ A JOURNEY TO EXCELLENCE

We finished the last chapter talking about looking good. The entire concept of cosmetic dentistry goes far beyond good. It is about being excellent. The instructions in this chapter will help to set an excellent dental office apart from a very good dental office.

Whether the patient accepts cosmetic treatment depends greatly on the dental team's ability to communicate, educate, and listen to the patient. "Inform before you perform" is a great office motto. Patients' dental I.Q. must be raised before, during, and after placement of cosmetic restorations.

Cosmetic dentistry is often an elective procedure. Dental insurance typically does not cover this treatment. If it is covered by insurance, it is usually to a much lesser extent than any other dental procedure.

GOOD NEWS FIRST

Patients considering cosmetic treatment usually do not care about the step by step procedures to make their teeth beautiful. In fact, it might even discourage many patients before they mentally "buy" the treatment.

To successfully encourage patient acceptance, the beautiful result of the procedure must be emphasized. Explain to the patient what can be corrected, then show them before and after pictures of other patients that have had this treatment. If a team member has had cosmetic treatment done by your dentist, ask the team member to show the patient their results. Once this exam is completed and the recommended treatment is **mentally accepted** is the time to tell the patient what the procedure entails and what to expect.

PATIENTS' INPUT

Patients must weigh the positives with the negatives while considering any cosmetic treatment. For example, porcelain veneers typically take two weeks to get back from the laboratory. Temporaries usually are not done since they are difficult and time consuming to make and their esthetic appeal is quite limited. Teeth prepared for porcelain veneers feel rough, build up plaque easily, and do not look good. Therefore the patient who travels or is in a highly visible job will need to strategically plan the dental appointment at an appropriate time.

One of our patients, who is the director of a top modeling agency, recently chose to have direct resin veneers because she felt she could not afford the time or the "look" during the two-week period. But the key here is that she was totally aware of her options and she had the right to choose.

The following instruction sheets are provided to our patients to allow their cosmetic journey to be an enjoyable, worthwhile experience. A journey to excellence, in not only their care, but the information provided to them.

This instruction sheet should be given to the patient before cosmetic treatment begins, so the patient can totally understand what is expected of them.

CARING FOR YOUR COSMETIC RESTORATIONS

Please follow these suggestions so that your restorations will look great for the maximum length of time possible.

1. Do not chew ice.

2. Brush with a ultra-soft toothbrush at least three times a day. Floss once a day, preferably at nighttime.

3. Do not use baking soda or any abrasive toothpastes. The toothpastes we recommend are: EpiSmile or Rembrandt

4. Do not rinse with mouthwashes high in alcohol. Alcohol softens bonding and weakens the bond of porcelain. For periodontal problems mix 1/2 water and 1/2 hydrogen peroxide. Rinse for one minute.

5. Sodium fluoride is the only home fluoride that should be used. Stannous fluoride or acidulated phosphate fluoride are not recommended for bonding or porcelain. Act fluoride is recommended for home use.

6. If you grind your teeth, consider having a custom bite guard made. This helps to avoid fracturing restorations while you sleep. If it is made with a soft material, it will also protect your teeth during any contact sports.

7. Do not pick at your restoration. You could pull open a small over-extension and shorten the life of your bonded or porcelain restoration. If you feel a rough edge with your

tongue, please call for an adjustment appointment to have the edge properly refinished.

8. Do not bite your fingernails or try to open bottles with your front teeth. The force can crack your new restorations.

9. To prevent staining, try to avoid or keep to a minimum coffee, tea, soy sauce, curry, colas, grape juice, blueberries or red wine.

10. To prevent fracture, avoid biting any hard foods with your front teeth such as: ribs, fried chicken, lamb chops, apples, carrots, hard rolls, etc. Avoid, or keep to a minimum, sugary foods as they can damage or weaken the bond between your veneer and tooth.

11. If your restoration chips or fractures, we will repair it at no charge for the first year. This applies to "no fault" accidents. It does not cover abuse, auto or sports accidents, or lack of compliance with the previous instructions.

KEEP SMILING AND THANK YOU FOR LETTING US TREAT YOU!

NEXT APPOINTMENT:

DAY:_____MONTH:_____YEAR:_____TIME:_____

This is given to patients after the cosmetic consult. This should be signed after the patient reads the information.

YOUR PORCELAIN VENEERS:

FROM PREPARATION TO SEATING

Congratulations! You are about to receive the most up-to-date, state-of-the-art veneer dentistry has to offer.

Porcelain veneers look very natural and are much more stain resistant than any other dental material. To keep your appointments running smoothly, the following is a summary of what you can expect:

Appointment 1: SCALING AND POLISHING

A thorough cleaning of all your teeth is necessary to ensure your teeth are free of plaque and tartar, and your gums are healthy.

Appointment 2: THE PREPARATION APPOINTMENT

Allow about 10 to 20 minutes per tooth for this appointment. Example: If you are having ten teeth prepared, allow two and 1/2 to three hours. Approximately 0.5 millimeter of enamel will be removed from the front of each tooth. This small amount of enamel is removed so that your veneers will not be bulky and will not cause any gum irritation.

Next, an impression is taken of your prepared teeth to send to the laboratory.

After this appointment, your teeth will feel slightly thinner and a little bit rough. It is not unusual for patients to say "they cannot get their teeth clean" during this period. The prepared teeth are not as smooth and might feel like a film of plaque has been left. Your tongue will notice the difference much more than your eye. Your teeth will not look noticeably different than before the preparation began.

Appointment 3: CEMENTATION

The cementation appointment is normally about two weeks after the preparation appointment. Allow twenty to thirty minutes for each tooth. The veneers are evaluated for perfect fit and color. Then the veneers are "bonded" to your teeth with a resin cement.

Your new "teeth" might feel a little thick or long the first couple of days. The thickness is due to the contrast of the thin teeth you have had for the last several weeks. The veneers plus the prepared teeth are usually about the same thickness as your original teeth. If your teeth are lengthened, it may take a couple of days before your teeth feel natural again.

Appointment 4: ADJUSTMENT

The adjustment appointment is about one week to ten days after the cementation. Final touch-ups and contouring are made on your veneers. This usually takes about 30 minutes.

SIGNED _____

 DATE_____

NEXT APPOINTMENT:

DAY:_____MONTH_____YEAR_____TIME_____

◆ ◆ ◆

MAINTAINING
ESTHETIC
RESTORATIONS
CHAPTER 11
PAGE 104

This sheet explains what is expected for Direct Resin Veneers.

DIRECT BONDED VENEERS:

WHAT TO EXPECT STEP BY STEP

Congratulations! Bonded Veneers are one of the most beautiful restorations dentistry has to offer!

The following is a summary of what to expect during your dental visits.

APPOINTMENT 1: SCALING AND POLISHING

It is very important that your teeth are thoroughly cleaned before the bonded veneers are placed. Your teeth should be free of plaque and tartar. This will also ensure your gum tissue is at its maximum health, which is very important for the continued success of the veneer.

APPOINTMENT 2: BONDED VENEER PLACEMENT

This appointment should be about one week to ten days after the scaling and polishing to ensure maximum health of your gums and teeth.

Allow approximately 30-60 minutes per tooth for the placement of the bonded veneers. Example: If you are having eight teeth restored, allow four-eight hours.

Bonding is a paste-like material that is applied to your teeth and sculpted with precision and artistry. After the tooth is the correct color and shape, a special light is placed over the bonding for about one minute. The light hardens the material so that it is like enamel.

After all of the bonded veneers have been placed and hardened with the light, they are contoured and polished.

Next, the veneers are made very shiny by a special polishing paste for bonded veneers.

Your new veneers can sometimes feel thick or long to your tongue. This takes a couple of days to feel natural again. They can be adjusted at your contouring appointment.

APPOINTMENT 3: CONTOURING AND ADJUSTING

This appointment is usually about one week to ten days after the veneers are placed. Thirty minutes should be allowed for this appointment.

Final contouring and smoothing rough edges or anything that feels "funny" to your tongue will be taken care of at this appointment. If contouring is necessary, another polish will be applied to the veneer to ensure it is as smooth as possible.

SIGNED _____

DATE_____

NEXT APPOINTMENT

DAY:_____MONTH:_____ YEAR:_____TIME:_____

"The Greatest Enemy of Excellence is Good."
Zig Ziglar

To special order your office instructions,
 call Reality Publishing Company at 1-800-627-5160.

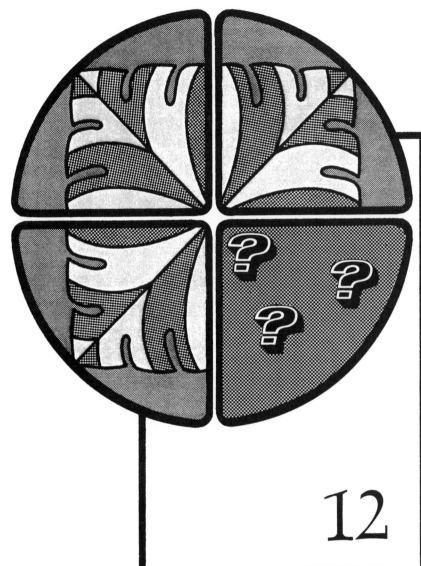

12
QUESTIONS
AND
ANSWERS

MAINTAINING
ESTHETIC
RESTORATIONS
CHAPTER 12
PAGE 108

CHAPTER 12

◆ QUESTIONS AND ANSWERS

The following questions were submitted by Lynn Lyons, R.D.H. and Linda Nash, R.D.H.

Q. *Since my teeth are now bonded, can I continue to use the fluoride rinse that my periodontist prescribed for me?*

A. It depends what type of fluoride your periodontist prescribed. Sodium Fluoride is the only safe fluoride for bonding. Stannous Fluoride is a slight etchant and incorporates a tin ion, which can cause staining. APF can etch bonding and porcelain which results in ditching and scratching, accelerating staining and wear.

Q. *Can the hygienist continue to use Nupro paste with fluoride on bonded restorations?*

A. According to Johnson and Johnson, there are four different grades of Nupro polishing paste. This rating system is based on the coarseness of the paste compared to enamel. The Nupro Plus is very coarse with a grade of 30+. Coarse Nupro is 27.5. Medium is 19.7. (All three of these grades contain pumice.) Fine rates 7.3, but does not contain pumice. The Fine Nupro is made of diatomaceous earth. We are not aware of any studies that have been completed testing the effect of diatomaceous earth on bonding or porcelain. I would recommend sticking with aluminum oxide for bonding and diamond polishing paste for porcelain until further studies have been done.

Q: *Will prophy paste such as Nupro damage the finish of porcelain or bonding? If so what is the damage and is it reversible?*

A: Nupro prophy paste can roughen bonding and scratch the glaze of porcelain. Most of the damage, however, is reversible if caught in time. See Chapter 9. "How to Deal with Deposits on Composites", to find out how to reverse the effects of prophy paste on bonding. For porcelain, see Chapter 5.

Q: *It is accepted by all dental hygienists that fluoride is beneficial to the enamel surfaces of teeth, but is that true for the surfaces of porcelain?*

A: The surfaces of porcelain do not benefit from fluoride. Acidulated phosphate fluoride (APF) has been found to etch glass and, as a result, it is not recommended for porcelain or hybrid composites which are filled with glass.

Q: *When the margins of bonding become stained, is there a procedure that can be done to correct the appearance without having to replace the entire restoration?*

A: Yes, there is a procedure that is very tedious and time consuming but it works, without further damaging the restoration. (See chapter 9.)

Q: *Are tobacco, coffee, and tea detrimental to the life of porcelain or bonded restorations?*

A: These substances can stain and possibly shorten the esthetic life of cosmetic restorations.

Q: *Can I use any toothpastes on my cosmetic restorations?*

A: Rembrandt or EpiSmile are two toothpastes specially formulated for cosmetic restorations. Baking soda is very abrasive, even to enamel. Tartar Control Crest is less abrasive than Colgate or Advanced Formula Crest, according to a than Colgate or Advanced Formula Crest, according to a recent study. However, another recent study totally contradicts the former's findings. Stay tuned. For now, we would recommend toothpastes specially formulated for cosmetic restorations.

Q: *Is at-home fluoride safe to use on porcelain restorations? And is there a possibility of using too much fluoride?*

A: The only safe home fluoride is sodium fluoride. Stannous fluoride releases a tin ion which could possibly stain the resin cement around the porcelain veneer. It has been documented in a recent study that frequent uses of sodium fluoride enhances the effect on inhibiting caries formation and remineralization of enamel. Porcelain would be unaffected by an increase in sodium fluoride rinsing.

Q: *Is a soft toothbrush still advisable for cleaning the restored surfaces of teeth?*

A: A soft or ultra-soft toothbrush is highly recommended for the restored surfaces to avoid scratching the surface.

Q. *Can the patient continue using Arm & Hammer Baking Soda to remove stain? Will it scratch the bonding?*

A. Baking soda is one of the most abrasive toothpastes and/or polishers around. It is an excellent stain remover, but even for enamel, we try to persuade patients to use it sparingly. Baking soda can ditch and abrade the top layer of bonded restorations creating increased roughness that may lead to more stain. When the patient uses baking soda or comes to their hygienist who might use an air-abrasive instrument, the stain is removed. But this actually creates a vicious cycle.

Ultimately the bonding is worn very thin and the projected esthetic and restorative life has also been shortened.

Q. *Can the patient rinse with hydrogen peroxide mouthwash?*

A. To date, there has been no evidence of hydrogen peroxide harming porcelain or bonding. As previously mentioned, hydrogen peroxide is recommended for patients with periodontal problems and cosmetic restorations. The American Academy of Periodontology has found hydrogen peroxide is nothing more than a placebo. Of course, there is other literature that states the opposite. Since hydrogen peroxide cannot stain or harm cosmetic restorations, we recommend it for patients with periodontal problems. In addition, we irrigate periodontal pockets with chlorhexidine. But don't prescribe it for home use - it will badly stain enamel, cementum, composite and porcelain restorations.

Q. *Can the hygienist use the ultrasonic or sonic scaler?*

A. These instruments are fine on enamel. However, on porcelain, they can scratch the glaze. This can accelerate staining and cause dullness of the finish. The bond of porcelain to tooth structure can also be affected by an ultrasonic or sonic scaler, possibly even weakening or breaking it. In addition, resin cement can be abraded by an ultrasonic or sonic scaler.

Composites can be harmed by an ultrasonic scaler by breaking out the filler particles or scratching and ditching the restoration. This could accelerate staining and microleakage, as well as shorten the esthetic life of composites.

◆ ◆ ◆

MAINTAINING
ESTHETIC
RESTORATIONS
CHAPTER 12
PAGE 112

Q. *Can the hygienist use the air polisher?*

A. Yes, an air-abrasive instrument is wonderful in preparing occlusal surfaces for fissure sealants and on virgin enamel. It is also very effective in thoroughly cleaning plaque from root surfaces, especially for patients with gingivitis and periodontal pocketing. However, this instrument has been shown to harm composite restorations by ditching and scratching the top layer.

An air polisher damages porcelain by scratching the glaze, causing dullness of the porcelain. Air polishers can also harm the resin bond around porcelain by ditching and scratching the resin cement, and perhaps even weakening the bond of the porcelain. Do not stop using this instrument, just use it selectively.

Q. *How long can the patient expect bonding to last?*

A. The expected life of bonded veneers and composite restorations in the oral cavity is contingent upon many variables. These variables include the operator, the material, the hygiene level of the patient, regular cleanings, pH of the particular patient's mouth, smoking, and bruxing.

Taking the ideal situation:

* least porous material

* perfect placement technique

* non-smoking patient

* perfect hygiene

* regular cleanings with aluminum oxide paste

the projected longevity can be seven to ten years for the esthetic and functioning life of composite restoration.

On the other end of the spectrum, taking the worst possible scenario, probably one to three years for the esthetic life, with the maximum of three for the functioning life.

Q: *How long will porcelain veneers last?*

A: No one really knows. The projected longevity is hard to determine because porcelain veneers are still quite new. With proper care, placement, maintenance program, etc, it would be safe to estimate 10 to 15 years.

Q: *If a patient has porcelain on tooth #8 and bonding on tooth #9, can I use one polishing paste?*

A: Yes. You can use aluminum oxide on both. Aluminum oxide does not harm porcelain, unlike diamond polishing paste, which can harm composites. But for the best results, use aluminum oxide paste on bonding and diamond paste on porcelain.

Q: *If the patient with cosmetic restorations is cavity-prone, I feel uncomfortable just using sodium fluoride at recall time. Do you have any suggestions?*

A: Yes. A daily rinse of any good sodium fluoride home rinse is recommended and has been shown, when used frequently, to be as effective as occasional use of APF. Rembrandt toothpaste also has fluoride. Advise the patient to watch their diet and home care. Explain to them the compromise that is necessary to ensure their entire oral cavity is healthy, esthetically and functionally. They will appreciate your care and the education will probably make their home care an A+.

"Only the Educated Are Free" — Epectitus

Have a particular or unusual problem polishing cosmetic restorations? Send your questions and a stamped, self addressed envelope to:

Lynn Miller, R.D.H.
11757 Katy Freeway, Suite 200
Houston, Texas 77079

INDEX

◆ ◆ ◆

MAINTAINING
ESTHETIC
RESTORATIONS
INDEX
PAGE 116

Porcelain , 4, 6-8, 12, 19-21, 23-26, 31-35, 39-48, 51, 53, 56-58, 77, 83-86, 88, 92-93, 95, 100-101, 109-114
Proximal , 22, 41, 43-45, 54-55, 58-59, 70, 78, 87-88

Resin
   Cement , 40-43, 46, 51-52, 54-55, 83, 92, 95, 104, 111-113
   Composite , 5-6, 20-21, 26, 88, 92-93, 96
   Direct , 6, 20-22, 31, 34, 53, 56, 75, 100, 105
   Indirect , 20-22, 31, 34, 53, 56, 75
Restoration
   Anterior , 7, 51, 53, 56, 84
   Composite , 20, 48, 51-52, 58, 77, 85, 91-92, 95, 112-113
   Porcelain , 4, 6-7, 42-43, 47-48, 53, 83-85, 92, 101, 110, 112
   Posterior , 7, 53, 56-57, 84-86
   Resin , 43, 53-54
Rubber Dam , 47-48

Scaler , 20, 43, 46-47, 58, 61, 65, 84, 96, 112
Small Particle , 56-57, 65, 87
Sonic , 46, 61, 84, 96, 112
Stain
   Extrinsic , 77
   Intrinsic , 77
Strips
   Aluminum Oxide , 71, 84, 87

Toothpastes , 20, 42, 65, 79, 92-93, 95-96, 101, 111, 114

Veneer
   Direct , 53, 56, 100, 105
   Indirect , 53, 56
   Porcelain , 6, 34, 39-40, 46-47, 51, 58, 93, 95, 100, 103, 111, 113
   Resin , 40-42, 51, 77